The
Gentle Subversive

New Narratives in American History

Series Editors
James West Davidson
Michael B. Stoff

The Gentle Subversive

RACHEL CARSON, *SILENT SPRING*,
AND THE RISE OF THE
ENVIRONMENTAL MOVEMENT

MARK HAMILTON LYTLE

NEW YORK   OXFORD
OXFORD UNIVERSITY PRESS
2007

Oxford University Press, Inc., publishes works that further Oxford University's
objective of excellence in research, scholarship, and education.

Oxford New York
Auckland   Cape Town   Dar es Salaam   Hong Kong   Karachi
Kuala Lumpur   Madrid   Melbourne   Mexico City   Nairobi
New Delhi   Shanghai   Taipei   Toronto

With offices in
Argentina   Austria   Brazil   Chile   Czech Republic   France   Greece
Guatemala   Hungary   Italy   Japan   Poland   Portugal   Singapore
South Korea   Switzerland   Thailand   Turkey   Ukraine   Vietnam

Published by Oxford University Press, Inc.
198 Madison Avenue, New York, New York 10016
http://www.oup.com

Oxford is a registered trademark of Oxford University Press

Library of Congress Cataloging-in-Publication Data

Lytle, Mark H.
  The gentle subversive : Rachel Carson, Silent Spring, and the rise of the
environmental movement / Mark Hamilton Lytle.
      p. cm. — (New narratives in American history)
  Includes bibliographical references and index.
  ISBN-13: 978-0-19-517246-1—ISBN-13: 978-0-19-517247-8 (pbk. : alk. paper)
  ISBN 0-19-517246-9—ISBN 0-19-517247-7 (pbk. : alk. paper)
    1. Carson, Rachel, 1907–1964. 2. Biologists—United States—Biography. 3.
Environmentalists—United States—Biography. 4. Science writers—United
States—Biography. I. Title.
QH31.C33L98 2007
570.92—dc22
[B]   2006049350

Printing number: 9 8 7 6 5 4 3 2 1

Printed in the United States of America
on acid-free paper

CONTENTS

FOREWORD

"There is properly no history; only biography," Ralph Waldo Emerson once observed. Few historians would agree. The frame of biography is simply too narrow to suit their aims. Measuring as they do the lines of change and continuity, concerned as they must be in reconstructing context, historians often find a single life unequal to these tasks. Yet some lives can serve such purposes if they are recreated with a telling sense of time and place.

Such surely is the case with Mark Lytle's *The Gentle Subversive*. In it, Lytle chronicles the life of Rachel Carson, a modern-day naturalist who produced one of the most important books of the twentieth century. When it appeared in 1962, *Silent Spring* sounded an alarm that still rings today, of an impending crisis brought on by the indiscriminate use of pesticides. What began as a well-intentioned effort to rid the environment of insects that destroyed crops and spread disease ended by threatening to poison the planet.

To sound such an alarm in the mid-twentieth century was to subvert a powerful paradigm, one that promised better living through chemistry and encouraged human hubris toward the natural world. It was also to storm the male-dominated bastions of business and science, which together touted these chemical nostrums. Carson mounted her assault by relying on her training as a scientist and her talent as a writer, and Lytle shrewdly links each chapter to a writing project—an essay or a book—whose twin tales of life and work reveal season by season the natural rhythms of her career. And more—for the story of this gentle subversive mirrors that of many women forced to operate in a male-dominated world. Carson's network of family, friends, and colleagues sustained her as she struggled with the critics arrayed against her from without and the terrible illnesses that beset her from within and eventually took her life.

It was a life more resonant than most, which succeeded in raising the environmental consciousness of the nation and, in a fashion, spawning the environmental movement itself. Hers is also a life as interdependent with those around her as the living things she so artfully portrayed are interdependent with one another. In that sense, *The Gentle Subversive* can be termed a truly ecological biography, one that is always sensitive to context and relationships. It fulfills one of the primary purposes of Oxford's New Narratives in American History series—to tell a story that makes its subjects come alive and makes historical sense of them.

James West Davidson
Michael B. Stoff
Series Editors

ACKNOWLEDGMENTS

Once upon a time, in a graduate seminar at Yale University, I asked the student sitting next to me what he was working on. "Environmental ideas in the nineteenth century," he told me. His project struck me as rather novel and unexpected, since in 1968 we had not yet celebrated Earth Day and few people thought much about the environment, especially as an academic subject. I myself had chosen a more conventional path toward diplomatic history and the Cold War but could not help thinking that this fellow, Donald Worster, was onto something, and indeed he was. His has been one of the most inspiring and enlightening voices in my generation of historians. Along with William Cronon, Carolyn Merchant, Richard White, and others, Don has helped to create a field that did not exist when I entered the profession. These historians have provided me with the intellectual resources I have needed to make a transition into a subfield some call "environmental diplomacy." Kurk Dorsey, who kindly read this book in manuscript, has been a welcome fellow traveler.

Several generations of Bard College students have made this journey with me. Among the many courses I teach at Bard, none has given me as much personal or pedagogical satisfaction as American Environmental History. Students in this course are the most intellectually curious and socially committed I teach. My colleagues in the Environmental Studies Program have been active in furthering my education. Daniel Berthold has been a rich source of ideas and readings. Bill Maple has done the best he can to drive ignorance from my mind, and Kris Feder has worked to convert me to her passion for Henry George. Much of the book was written during a year I spent at University College Dublin. My good friends and colleagues Michael Laffin and Ronan Fanning made the year both possible and rewarding. Richard, Kathy, and Lizzie Aldous made sure I felt at home. Whether introducing me to the Ipswich Town "Tractor Boys," providing a venue for the Super Bowl, or discussing the world of history from the ins and outs of departmental politics to the dynamics of the international system, Richard was a constant source of fellowship and support for my writing projects.

Librarians at Bard, and especially at the Beinecke Library at Yale University, have made the research for this book a pleasure. In that regard, I would like to pay a special tribute to Linda Lear, whose work did so much to guide mine. Linda provided timely advice and led me to essential resources. My research assistant, Nick Buccelli, appreciated exactly the kind of material I was seeking. Dr. Stanley Freeman, Jr., made available photographs from his collection so that the book could keep Carson's "spirit alive for a new generation." Mark Stoll, an early reader of the manuscript, shared some of the work he is doing on Rachel Carson. He also introduced me to Fritz Davis, whose historical sleuthing established a most significant link between Carson and the work of Charles Elton. Fritz generously shared his insights with me. Several other anonymous readers were equally helpful and incisive. Rob Anderson, a dedicated environmentalist and birder, read the manuscript in its early form. So did my son Jesse, who, despite his trip to the "dark side" as an administrator at Mount Holyoke College, has preserved his keen critical judgment. I'm grateful to both for providing me with another generation's point of view. My wife, Gretchen, has lived patiently through my writing process, as always constructively skeptical and supportive.

This book is dedicated to Jim Davidson and Mike Stoff for good reason. Jim and I set out long ago in writing *After the Fact* as the "Mutt and Jeff" of history—Jim is very tall. Along with Mike we formed what I think of as the "Three Amigos" of history. We have had a lifetime conversation about what history should be and, even more, how it should be written. History may be a narrative form, but few historians actually write narrative. Thus, when Jim and Mike conceived the New Narratives in American History series, I was eager to be part of it. The idea to make Rachel Carson my subject was my own; the finished book involved heavy collaboration. On our morning jogs, Jim and I have helped each other think about our respective stories (his is a forthcoming biography of the antilynching crusader Ida B. Wells). Mike took my initial draft and showed me how to shape it into a real narrative. To have such thoughtful and inspired editorial input is rare for most any author.

Finally, I thank Rachel Carson for being such an inspirational writer.

PROLOGUE

RACHEL CARSON COULD NOT WAIT TO SHARE THE NEWS WITH Dorothy. The news was so exciting that Dorothy Freeman had to be told, for she had become specially important in the network of personal and professional relationships, many of them with women, that Rachel had been building for years. Over the years, these reciprocal relationships provided Rachel with sustenance and support in much the way that the natural world about which she wrote so poetically sustained its constituent parts in the interdependent ecology of life. In a strange and very personal fashion, Rachel Carson had fashioned an interdependent ecology all her own, and Dorothy had become a vital part of it.

Now, as spring turned to summer in 1958, Rachel's new book project took her to New York in June for a meeting with William Shawn, the legendary editor of *The New Yorker* magazine. The meeting had gone even better than she hoped. Shawn gave her almost three hours of his precious time and told her he wanted as many as 50,000 words for an article about the threat that indiscriminate pesticide spraying posed to both wildlife and humans.

He understood how hard it would be for her to maintain the sober tone of a disinterested scientist, and he didn't much care. "After all," Shawn told her, "there are some things one doesn't have to be objective and unbiased about—one doesn't condone murder!" That meant at least three installments in *The New Yorker* and a huge boost for her project, a book whose working title was *Man Against the Earth.*[1]

Rachel could barely contain herself and hoped Dorothy would share her joy, but she knew her friend had reservations about "the poison book." Dorothy told Rachel that an investigation into a subject as morbid as pesticides could not possibly reflect her gift for bringing the beauties of nature to the printed page. Still, for Rachel, there was no turning back. The book already consumed much of her time and energy. Every day she was corresponding with a widening network of scientists, government officials, and concerned citizens. A Long Island lawsuit over pesticide spraying was moving forward, and the plaintiff's evidence was helping Rachel build her own case. That meant fewer letters and less time for Dorothy. And sometimes when she did write, Rachel just poured out what was on her mind without her usual sensitivity to Dorothy's feelings.

Dorothy was something of a latecomer to Rachel's circle of friends. Rachel first met her in the summer of 1953 after building a cottage on rugged Southport Island along the coast of Maine. The sense of wonder that infused Rachel's earlier books about the sea inspired their friendship. They loved nothing more than walking in the woods or exploring the crevices in the rocks lining Southport's shore. And despite Dorothy's reservations about this new project, her letters brightened even the darkest days as

Rachel Carson (right) with her Maine coast neighbors, Dorothy and Stan Freeman. The three shared a love of the sea and its creatures. (SOURCE: Stanley Freeman, Jr. Permission granted by the Freeman Family Collection.)

Rachel struggled with her investigation of the pesticide DDT and its impact on the environment.

In the past, they had filled their letters with news of plants in bloom or their desire to spend time together spotting birds or even the occasional whale in the bay. For her part, Rachel hoped Dorothy would come to understand how much she believed in what she was doing. "There would be no future peace for me if I kept silent," she explained. "I wish I could feel that you want me to do it." Given what this project might mean to future generations, Rachel believed she had to press on even at the risk of being called a crank or, even worse, a subversive.[2]

True enough, Dorothy did find pesticide abuse a grim subject in contrast to the celebration of nature that had inspired Rachel's earlier books. That was not, however, the only source of her discomfort. In her last book, *The Edge of the Sea*, Rachel had written about the seashore world they explored together. What Rachel was calling *Man Against the Earth* was far more technical and, to be honest, downright frightening. Unlike Rachel, Dorothy was no scientist and frankly found the research into "chlorinated-hydrocarbon molecules" not nearly as interesting or uplifting as the shorebirds and sea creatures of Rachel's earlier books.

Dorothy also knew how quickly Rachel could become overwhelmed when she took up a new project. Added to the burden of caring for her adopted grandnephew Roger and her mother, Rachel was immersed in a potentially explosive subject and preoccupied with gathering information. Dorothy feared what the additional pressure might do to Rachel. How would she hold up under what could be sulfurous attacks from angry scientists, business executives, and government bureaucrats? Her friend was at heart a gentle soul, with little taste for the public spotlight and the enervating combat her new book might well provoke. Nagging health problems already sapped her energy. Rachel admitted to Dorothy that her "disease" (most likely an undiagnosed ulcer) was "no doubt aggravated by the nature of my subject" as well as "by the fierce drive to get it done."[3]

During the summer of 1958 Carson took a brief vacation from her research. She and Dorothy managed to soothe the hurt feelings between them. "That awful weight has lifted," Rachel wrote at the end of August. She returned to Maryland and her work that fall with renewed energy: "I'm actually excited and happy about the book. For all its unpleasant features, it is something with so many complexities, that it is mentally stimulating, as are the contacts with many of the people I talk or write to."[4]

Those contacts were essential to the way Rachel approached each new book. As a trained biologist who spent years translating jargon-filled government reports into English, Carson knew how to extract relevant materials from dense data. She did not, however, simply find her subject and consult relevant articles and experts. Rather, she constructed networks of correspondents with whom she raised questions and from whom she sought explanations. That was the strategy she was now employing with *Man Against the Earth*. By 1958 her research files began to expand into what would become a small archive filled with letters and articles from hundreds of sources. Most of her correspondents supported her project. They shared her concerns for protecting the natural environment and appreciated her gift for translating their scientific jargon into language any reader could understand. To guarantee that her translation of their work did not misrepresent their views, she sent relevant passages to them for editing and verification.

This approach created an intellectual ecology for her books. Just as ecology stressed the interconnectedness of living things, so Carson took a broad, synthetic, mutually dependent approach in her research. To that end, she brought together the work of scientists whose writing was often so specialized that it seldom appeared in the same journals and so impermeable that it seldom reached the public. Her respect for ideas and her ability to convey them gained the confidence of scientists and turned many of them into champions of her books.

Rachel knew such scientific support would be vital to the acceptance of *Man Against the Earth,* especially given the controversial nature of its subject. Controversy was also much on the mind of her literary agent, Marie Rodell. Rodell, like Carson, wanted the book to reach the widest audience possible. She knew Rachel was taking great pains to be sure that her data were both persuasive and accurate. Rodell also understood that women received scant respect from male scientists, who dominated her field. Carson, after all, spent her career as a government scientist, not in the role of aquatic biologist for which she trained, but as a publisher for her largely male colleagues. Still, Rodell, ever the practical literary agent, did not want a general audience driven away by turgid discussion of technical points. "The scientific word should be avoided whenever possible," she advised.[5]

It was with a general audience in mind that Rodell wanted to settle the question of the title. Several of those they considered, such as *The War Against Nature* and *At War with Nature,* struck Rodell as too adversarial. An intemperate attack on the chemical companies, she feared, might trigger a backlash that would give critics "the opportunity to yell 'crank.'" Reading the draft gave her an inspiration. The chapter on birds, entitled "Silent Spring," struck Rodell as so beautifully written that she wondered "if 'Silent Spring' mightn't make a title for the whole book?"[6]

The idea appealed to Carson. Spring had always been a special season in her life, when fields turned green with new life, flowers burst into bloom, and bird songs filled the air. The idea that poison would end that life and silence the music filled her with dread. To protect the natural world that inspired her earlier books, Carson believed she must warn against what she saw as impending disaster. She could not be silent even if the men of science, many of them smug experts in white lab coats who promised "better living through chemistry," dismissed her warnings as feminine hysteria. The dead songbirds, victims of pesticide sprays that her friend Olga Huckins had discovered in her backyard, offered mute testimony to dangers that were all too real.

Olga Huckins, like Dorothy Freeman and Marie Rodell, was part of that network of women who admired Carson's writing and supported her work. As a former literary editor at the *Boston Post*, Huckins too had been a woman in a man's world. She and Carson began corresponding in 1951 and shared an interest in nature and birds, as well as writing. Both were involved with the Audubon Society, a national network of local bird watching groups dedicated to conserving wildlife and habitat. Over time Audubon members like Huckins adopted an ecological vision— they understood that humans and nature were linked in a common living system. Pesticides devastating the birds in their backyards, they now believed, also posed a threat to human health.

While the national leadership of the Audubon Society was largely male, its membership was heavily female. That was even more the case for the national networks of humane societies and garden clubs that became another critical source of support for Carson. These were women for whom the well-being of animals and the health of nature was a matter of daily concern. Carson gave voice to their fear that "the indiscriminate use of biocides in the environment, soil, water, air, and food" endangered the living systems on which all life depended.[7]

Over the next year Carson made great strides on the project, but her subject grew more and more complex. She had always seen herself as a naturalist and a writer who in her books revealed the wonders of nature and the ecological interdependence of the living world. Indeed, her first book, *Under the Sea-Wind* (1941), bore the subtitle *A Naturalist's Picture of Ocean Life.* She elaborated those themes in her huge bestseller *The Sea Around Us* (1951) and again in *The Edge of the Sea* (1955). In the atomic age she grew ever more uncomfortable with the power of science and technology to undermine or even destroy that ecological interdependence. Pesticides were yet another lethal weapon that threatened nature and the ecological systems on which human life depended.

Carson saw that the book she began as a defense of nature against human assault was evolving into a vital lesson in applied ecology. In disturbing the balance of nature, pesticides posed every bit as great a threat to humans as they did to other living things. In her initial outline for *Silent Spring,* she anticipated that just one of twelve chapters would focus on the impact of pesticides on human health. But the more she learned about the health dangers, the more her moral outrage grew. One chapter turned into four in the final draft. William Shawn caught her impassioned tone when he likened indiscriminate spraying to "murder." Carson was persuaded that many experts either failed to recognize or chose to ignore the potential hazards of pesticides. She was convinced that the weight of her scientific evidence would defeat the skeptics among them. And once the public had the necessary information, citizens could make informed decisions about what Carson believed was a matter of life and death.[8]

Perhaps a different kind of person sharing Carson's ecological vision and moral outrage would have publicized her concerns in lectures and newspaper articles and on television and radio. All those venues would have furnished a forum for her ideas. But the intensely private Carson never felt comfortable on display. She was and always had been a writer. She produced her first book before she was ten years old and published her first article at the age of eleven. The problem she now faced was to determine just what kind of book *Silent Spring* was going to be. But the human-health issues made her research more complicated, her case more controversial, and the date for a finished manuscript more uncertain.

Yet, just as she sought to marshal her energies to finish the book, her own health failed her. In January 1960, she learned she had developed a duodenal ulcer. "I should think [the ulcer] would have waited until the book was done," she told her editor, Paul Brooks. "I'm sure some people would think that the subject matter I'm dealing with is the cause of the ulcer, but I'm equally sure it isn't." Battles with pneumonia and recurring sinusitis followed. Somehow she managed to keep writing, and by March a first draft was emerging. The conviction that she had a vital message to deliver drove her on. Then she discovered two new cysts in her left breast. Twice before, such tumors had appeared but had turned out to be benign. This time, her doctor suspected cancer and, as a precaution, performed a radical mastectomy. When Carson asked about the pathology report, he reassured her that, though the cysts were cause for worry, her condition was only "bordering on malignancy." The surgery was painful and the recovery slow, but given her doctor's assurances, Carson returned to writing, though only for a few hours each day.[9]

When Dorothy came for a brief visit in late April, she was dismayed to see how much her friend suffered. A slender woman, Rachel always took care of her appearance. Now, her face was gaunt and her body emaciated. The following December Rachel discovered the truth—her doctor had lied. "The tumor was malignant, and there was even at the time evidence that it had metastasized," she learned.[10]

Carson faced the battle of her life. Subsequent X-ray treatment caused her a "serious diversion of time and capacity for work because there are some side effects," she confessed with typical understatement. The nausea, cramps, and near-numbing fatigue tormented her daily and sometimes left her bedridden or confined to a wheelchair. She was forced to wear a wig whenever she appeared in public. Despite her incapacity, she hoped "to work hard and productively. Perhaps even more than ever, I am eager to get the book done." Brooks, ignorant of how ill Carson was, wondered whether they might aim for publication in the summer of 1961. Carson responded with a definite no.[11]

So despite her ill health, Carson pushed on to finish *Silent Spring*. All the pieces seemed in place for her to do so. Two questions remained: Would the author live to publish it? And would the public heed her warning? Her network of scientific correspondents had supplied her with mountains of evidence with which to make her case. Friends and professional acquaintances urged her on and gave her the emotional support she needed. Her powers as a writer had never been greater. Those who knew anything about Carson—the passion of her convictions, the will of iron that supported them, and the eloquence of the prose that expressed them—believed the answer to both questions was yes.

Notes

1. Carson to Freeman, June 12, 1958, in Freeman, *Always Rachel,* p. 257. *Silent Spring* actually went through a number of working titles, including *Man Against Nature, The Control of Nature,* and *How to Balance Nature,* none of which ever satisfied Carson or her literary agent, Marie Rodell.
2. Carson to Freeman, June 28, 1958, in Freeman, *Always Rachel,* p. 258.
3. Ibid., p. 259.
4. Carson to Freeman, August 30, 1958, in Freeman, *Always Rachel,* p. 266.
5. Lear, *Rachel Carson,* p. 377.
6. Ibid. pp. 377–378; see also Brooks, *House of Life,* pp. 368–369.
7. Hazlett, "'Woman vs. Man vs. Bugs'" pp. 701–729; see also Norwood, *Made from This Earth,* pp. 143–171.
8. I am indebted to Mark Stoll for enhancing my understanding of the religious roots of Carson's moral outrage. Through her mother, Maria Carson, and her Presbyterian forebears she imbibed a reverence for nature, an antimaterialist ethic, and a reformist impulse. See, for example, Stoll, *Protestantism, Capitalism, and Nature in America,* and "Rachel Carson: Nature and the Presbyterian's Daughter," a draft chapter the author kindly provided me.

9. Carson to Brooks, March 16, 1960, Rachel Carson Papers, Beinecke Library, Yale University (hereafter cited as "RCP-BLYU"); Brooks, *House of Life*, pp. 270–272; Lear, *Rachel Carson,* pp. 367–369.

10. Carson to Brooks, December 27, 1960, RCP-BLYU.

11. Lear, *Rachel Carson*, pp. 377–385; see also Brooks to Carson, November 10, 1960, and Carson to Brooks, December 27, 1960, RCP-BLYU.

· One ·

SPRING

Sense of Wonder: *Under the Sea-Wind*

Sɪɴᴄᴇ ᴀɴᴄɪᴇɴᴛ ᴛɪᴍᴇs, ᴍᴀʀɪɴᴇʀs ʜᴀᴠᴇ ᴛᴏʟᴅ sᴛᴏʀɪᴇs ᴏꜰ ꜰᴀʙᴜ-lous sea creatures and ferocious monsters that lurked beneath the waves. Yet, few humans ever knew what manner of life the oceans might contain. Then, in November 1941, a remarkable book appeared—*Under the Sea-Wind* by Rachel Carson. One veteran voyager confessed to Carson that after a lifetime of travel on the seas he had been so much "impressed by [their] apparent lifeless aspect—'not a snout nor a spout did I see'" that he craved to know more about the mysteries the oceans held. *Under the Sea-Wind* satisfied his curiosity as it took him along the seacoast and below the water's surface, where snouts and spouts abounded.[1]

In a departure from traditional writing about the seas, Carson told her stories from the perspective of the creatures themselves: Silverbar, a sanderling whose migratory flight pattern stretched from the Arctic Circle to Patagonia at the southern end of Argentina; Scomber, a mackerel who followed his school from the shores of New England to the Continental Shelf, where it drops into the Atlantic's great chasm; and Anguilla, an American eel who journeyed to the Sargasso Sea, an ocean abyss south of Bermuda where American and European eels come thousands of miles to spawn. These stories not only immersed readers in the oceans' depths; they also reflected what scientists knew about the life cycles of sea creatures. Carson dramatized their urge to survive and reproduce in the eternal patterns of their species.

Glowing reviews poured in. The *New York Times* called it a "breathtaking canvas of the fierce struggle for life that exists along the shore, in the open sea, and along the sea bottom." Almost all who read *Under the Sea-Wind* agreed that Carson had managed to combine science and poetry. One critic considered the book so "skillfully written as to read like fiction" but then added that it was "in fact a scientifically accurate account of life in the ocean and along the ocean floor." Scientists were equally generous in their praise, admiring its poetic language and yet accurate science. Where, some wondered, had this gifted storyteller come from and how had she found her unique voice?[2]

As a young girl, Rachel Carson had two passions—writing and nature. Where the idea of being a writer came from remained a mystery, even to her. "I have no idea why," Carson later confessed. "There were no writers in the family." All the same, she remembered that she "read a great deal almost from infancy, and I suppose I must have realized someone wrote the books and thought it would be fun to make up stories, too." And she did, publishing her first one at the age of eleven. While the books she wrote contained stories, fact more than fiction informed her writing. Those facts were drawn largely from her study of nature: "I can remember no time when I wasn't interested in the out of doors and the whole world of nature." The source of that curiosity was no mystery at all. "Those interests, I know, I inherited from my mother and have always shared with her," Carson acknowledged. "I was rather a solitary child and spent a good deal of time in the woods and beside streams, learning the birds and the insects and the flowers."[3]

Rachel was the third of three children of Robert and Maria Carson. In some ways, she was almost an only child since her sister and brother were so much older. When she was born on May 27, 1907, her sister Marian was ten and her brother Robert eight. After Robert's birth, her father settled the family on a farm outside the village of Springdale, Pennsylvania, 15 miles up the Allegheny River, northeast of Pittsburgh. The setting evoked an earlier pioneering era. When the Carsons settled in Springdale, farms, woodlands, and streams surrounded the village. The farmhouse into which Rachel arrived had once been a log cabin. It had no central heating or indoor plumbing and only a few ceiling fixtures for electric light. Her father raised chickens and other farm animals and sold fruit from his apple and pear orchards. Maria Carson kept a large vegetable garden behind the kitchen.

Robert Carson had been attracted to the property, not so much for its rural charm but for its potential for real estate development. By the early twentieth century industrial Pittsburgh was rapidly expanding toward Springdale. Carson assumed his land would eventually make way for home sites, but in the end neither he nor Springdale ever achieved prosperity. Industrialization

Maria Carson and her children Marian, Rachel, and Robert. Rachel was the youngest of the three Carson children and the object of her mother's devotion. (SOURCE: Courtesy of the Rachel Carson Council.)

brought blue-collar workers, smoke-belching power plants, and an end to the community's bucolic setting. Rachel later remembered her hometown mostly for the stench of its glue factory and the drabness of its streets.

Robert Carson cobbled together a living by selling an occasional lot and traveling to sell insurance. To supplement his unsteady income, he eventually took a part-time job at the nearby West Penn Power Plant. His two older children attempted to escape the confines of home and small town living—Marian through a brief and disastrous marriage, Robert Jr. through a tour of service with the Army Air Corps in France during World War I. Both finally returned to Springdale and joined their father as employees of West Penn Power, one of the few opportunities in town for those who chose to live and work there.[4]

Maria Carson harbored higher hopes for Rachel, the child of her later years. Marian and Robert quit school after the tenth grade. Neither shared Maria's love of music, books, and, above all, the woods and fields that surrounded her home. Rachel, a "dear, plump, blue-eyed, little baby" born in Maria's thirty-eighth year, was different. She named Rachel for the person who had most shaped her own life, her mother, Rachel Andrews McLean. Like Maria, Rachel McLean had a disappointing marriage. Her husband Daniel, a Presbyterian minister, contracted tuberculosis at the age of 30. The disease left him periodically unable to serve in the pulpit. After ten years of struggle against the disease, he succumbed in 1880. The care and education of her two daughters, Ida and Maria, fell almost completely to Rachel, who welcomed the challenge. As a result, Maria grew up in a world of women defined by Presbyterian rectitude and the powerful personality of her own mother. This was the world into which Maria initiated her daughter Rachel. Rachel Carson, though comfortable with men, formed her most important friendships with women.[5]

With her other children in school and her husband frequently on the road selling insurance, Maria had all the time in the world to manage the life of her darling little daughter. (Save for a few years when Rachel went off to college, the two lived together until Maria died in 1958. Maria kept Rachel's house, edited and typed her manuscripts, and offered unflagging encouragement. Rachel, in return, provided for her mother, both financially and emotionally.) Maria's management began with her daughter's education. That role came naturally to a former teacher with time on her hands and an emotional hole at the center of her life. The natural world was for her a sanctuary, full of wonder and spiritual sustenance. Now it would become her classroom and Rachel's as well.

Rachel later recalled that she had been "happiest with wild birds and creatures as companions." Almost every day young Rachel and her mother headed outdoors to discover the living things in the fields and woods nearby. Maria taught Rachel to observe closely and to remember what she saw. When Rachel brought living specimens home to show her mother, Maria insisted that after studying them Rachel return them unharmed to their habitats. Those lessons stuck with Carson all her life.[6]

School did little to loosen the bond between mother and daughter. Rachel had to walk almost two miles to reach her classroom, but when the weather turned foul or sickness broke out, Maria kept Rachel at home. She feared the potentially fatal childhood diseases that lurked everywhere: scarlet fever, diphtheria, typhoid, and pneumonia. Days away from school allowed Rachel to do what she loved most, next to her nature studies, to read. For her, as for her mother, books became an escape from the restricting confines of Springdale. Early on she developed a particular taste for authors who wrote poems and stories about the sea. Herman Melville, Joseph Conrad, and Robert Louis Stevenson were among her favorites.

Despite frequent absences from school, or perhaps because of them, Rachel became an outstanding student. She worked diligently and from second to seventh grade received nothing but As save for an occasional B in penmanship. Friendship was another matter. Most of Rachel's classmates found her reserved and distant and Maria dauntingly stern. Maria Carson seldom entertained guests and certainly did not encourage Rachel to bring friends home. Both were undoubtedly a bit embarrassed by the shabby household, and each was in turn the other's best friend.[7]

Carson devoured books by the score and early on began to imagine herself as a writer. By age nine she produced her first books, including a gift to her father. "This little book I've made for you my dear," she wrote in her inscription, "I'll hope you like the pictures well; the animals you'll find in here—About them all—I'll tell." Each page contained a drawing of an animal with an accompanying verse.[8]

At age eleven Carson began sending out her stories for publication. Maria had introduced her to *St. Nicholas Magazine* for young readers. Under the editorship of Mary Mapes Dodge (author of *Hans Brinker; or, The Silver Skates*, 1865), *St. Nicholas* attracted well-known writers from America and England, including the likes of Mark Twain, Louisa May Alcott, Robert Louis Stevenson, Rudyard Kipling, and L. Frank Baum, author of *The Wonderful Wizard of Oz*. It also published young writers who would become part of the American literary canon—E. B. White (of *Stuart Little* fame), Edna St. Vincent Millay, William Faulkner, and F. Scott Fitzgerald.

In 1899, *St. Nicholas Magazine* introduced a feature in which it published work by and for children up to the age of eighteen. Rachel sent off her first piece in May 1918 just as she turned eleven. In "A Battle in the Clouds," she retold a story her brother Robert described in a letter home. His reports of a brave Canadian aviator made a powerful impression on Rachel. "The Germans," she wrote, ". . . could not but respect and admire the daring and courage of the aviator and did not fire until [his] plane safely landed." The editors at *St. Nicholas* rewarded the precocious voice with a silver badge for excellence. Three more stories on war-related themes followed within the year, including one, "A Message to the Front," that received a gold medal. Her fourth publication made her an "honor member" in the magazine's League and earned a $10 prize. Her career as a writer was under way.[9]

Like all writers, young Rachel had her disappointments. *St. Nicholas* turned down one piece but then asked to use it for "publicity work." The $3 she received at a penny per word marked her debut at age fourteen as a professional paid writer. Several months later, three magazines, including *St. Nicholas*, rejected "Just Dogs," a well-written story on a rather hackneyed subject. In the process, Rachel revealed a hardheaded business sense that would belie the notion that she was an "incurable romantic." Author's Press apparently sold instructional services to aspiring writers and profited from their desire to be published. In submitting "Just Dogs," Rachel laid out her own conditions:

> The accompanying manuscript is submitted under the terms on which I bought The Irving System, i.e. a typewritten manuscript will be examined and if possible, sold, free of charge. As I have use of a typewriter, it is more convenient for me to make use of these terms rather than those described in your letter; by which $2 is paid for the reading of the manuscript which is sold on a 10% commission basis.[10]

Author's Press probably would have taken her $2, but the editors refused to accept the story on her terms. Although it never dictated what she wrote, the quest for financial security drove Carson to publish throughout her adult life.

St. Nicholas did more than encourage Carson as a writer; it reinforced her budding career as a naturalist. Each issue contained a feature called "Nature and Science for Young Folks." In it, the magazine exposed its young readers to the ideas of the nature studies movement. That movement reflected the arcadian sentiment that flourished in early twentieth-century America and harkened back to the eighteenth-century parson-naturalist Gilbert White of Shelborne, England. In his widely popular and often reprinted book *The Natural History of Shelborne*, White depicted people living a simple life in harmony with other creatures. He made it his practice to balance his pastoral duties with long walks through his parish, closely observing the seasonal changes of its flora and fauna. White intended nothing more than to record as a "philosopher" of nature "the life and conversation of animals." Birds and their migrations were his special interest, yet as one observer remarked of him, White "grasped the complex unity of diversity that made Shelborne and its environs an ecological whole."[11]

In Shelborne, as he wandered along twisting paths and gently rolling fields, White could see the "balance of nature" that later came to constitute the core of Rachel Carson's ecological sensibility. Under the Creator's guiding hand, White concluded, nature "converts the recreation of one animal to the support of another." Every creature, no matter how insignificant or ugly it might seem, contributed to the natural cycles. "Earthworms, though in appearance a small and despicable link in the chain of nature, yet, if lost would make a lamentable chasm," he observed. The worms provided food for birds, which in turn became food for foxes and people. White blended his celebration of the natural world with his religious faith that a divine providence "had contrived this beautiful living whole."[12]

White's pious regard for the interdependence of nature was very much alive in the United States in the early twentieth century. Among his disciples were those educators and scientists who embraced the nature studies curriculum, a movement with links to the Reformed Presbyterian theology in which Maria Carson raised her daughter. Proponents of this program worried that the children who moved from farm to city would lose contact with nature. It was not enough to teach city students to collect and identify species and recognize the scientific order of the flora and fauna of woods and fields. Students needed direct contact with nature to sense its grandeur and mystery. Many nature studies educators feared that a curriculum limited to classroom science would destroy the love for nature the program sought to teach. One educator described science teaching at the elementary level as "a sort of military ration of a special few." By contrast, nature study sought to "create a passionate love for things in God's great school." Direct contact would lift young children out of the humdrum of everyday life and instill a higher spiritual and moral sense. Cornell biologist Anna Botsford Comstock, one of the program's staunchest advocates, reminded her colleagues that "nature study in the schoolroom is not a trouble, it is a sweet, fresh breath of air across the heat of radiators and the noisome odor of over-crowded small humanity." Her textbook *Handbook of Nature Study* taught that nature was a holy place and its study would reveal the grand design of the Creator. "If the earth is holy, then the things that grow out of the earth are also holy," one of her colleagues observed.[13]

Another of the movement's prominent apostles was Comstock's associate at Cornell, Liberty Hyde Bailey. "It is becoming more and more apparent," Bailey wrote in 1901, "that the ideal life is that which combines something of the social and intellectual advantages and physical comforts of the city with the inspiration and peaceful joys of the country." Camping, wilderness exploration, bird watching, picnicking, national parks, and scenic gardens were all avenues through which city dwellers could experience those "peaceful joys." Bailey believed that for city children summer camping "with firm but kindly discipline and something like Spartan fare" should be part of their education. To proselytize his ideas, he became editor of *Country Life in America*, the era's most popular suburban magazine. In it, Bailey promoted a vision of scenic landscape that would nurture "the spirit of pleasant inquiry, of intellectual enthusiasm, and of moral uplift."[14]

In their approach to nature studies, Comstock and Bailey described the education Maria Carson tried to impart to her children. As a reader of *St. Nicholas Magazine,* Maria imbibed many of their ideas well before Rachel was born. Marian and Robert Jr. used Comstock's *Nature Studies* readers in their classrooms. Now from Comstock's texts Maria adapted lessons she could teach to her children while using her farm and its woodsy environs as a classroom. And from Bailey she derived another idea that shaped Rachel's thinking. Where Gilbert White and many "back to nature" enthusiasts assumed that God had placed humans at the center of his creation and placed other living things for their use, Bailey understood from Charles Darwin that creation was an ecological process in which all living things played a role. "The living creation," Bailey noted, "is not exclusively man-centered; it is biocentric." Maria Carson inculcated that biocentric ideal in her children, and Rachel took it truly to heart.[15]

In July 1922, Rachel published an essay in *St. Nicholas Magazine,* "My Favorite Recreation," in which she described a day of "birds'-nesting" with her dog Pal. Carson knew enough about birds by age 14 to recognize their calls and the distinctive shape and materials of their nests. As she and Pal entered the woods, "majestic silence, interrupted only by the rustling breeze" surrounded them. Then, she heard the "cheery 'witch, witchery' of the Maryland yellow-throat." Half an hour of trailing led Rachel and Pal to a "sunny slope" where in "some low bushes, we found the nest containing four jewel-like eggs." The day's inventory included bobwhites, orioles, cuckoos, and hummingbirds; and then, to her joy, she heard a call that sounded like "Teacher, *teacher,* TEACHER," the distinctive voice of the ovenbird.

Nearby, she came upon its home in a "little round ball of grass, securely hidden on the ground." This was a part of the world she loved to observe but never disturbed.[16]

As Rachel approached her junior year in high school, her parents decided she had outgrown the meager resources of School Street School in Springdale. School Street had so few students it offered only tutorials rather than regular classes. Robert and Maria could not afford the cost of train fare to send her to a regular high school. But by 1923 Robert had taken his part-time position at West Penn Power, so he could afford the carfare Rachel needed to attend Parnassus High, a trolley car ride away. During her two years at Parnassus, she widened her circle of friends and participated in sports and the other extracurricular activities so central to the high school years. Classmates discovered in her a lighter and more irreverent side, though she nonetheless graduated first in her class. In the yearbook, her classmates captured her dedication to excellence:

> Rachel's like the mid-day sun
> Always very bright
> Never stops her studying
> Until she gets it right.[17]

Parnassus required all its graduates to write a senior essay. In hers, Carson revealed two salient qualities. From her grandmother McLean she inherited a stern Calvinist moralism. Superficiality, intellectual laziness, and moral indifference were qualities Rachel condemned. For her, the waste of one's intellectual gifts was akin to the "reckless squandering of natural resources." From the progressive conservation tradition of the early twentieth century Rachel also gained an appreciation of the power of applied intelligence to improve society. Just as progressives championed disinterested expertise in pursuit of the common good, Rachel preached the value of the independent pursuit of truth or, as she put it, "the use of our own reasoning power." What she lacked at this youthful age, however, was a sense of political engagement. On the issues that aroused many progressives and young idealists of the early twentieth century—poverty, corruption, and social injustice—Rachel was silent. Her recipe for human betterment called for individual striving rather than commitment to social movements.[18]

By the age of eighteen Rachel was ready for a world beyond the confines of Springdale. Maria Carson understood that college was essential if her daughter were to realize the ambitions that Maria had long nurtured. But to lose her intimate contact with Rachel was almost unthinkable. More practically and even more importantly, the cost for most colleges was beyond the Carsons' limited means. The solution to this seeming dilemma lay in nearby Pennsylvania College for Women (PCW), which has since become Chatham College. PCW's Pittsburgh campus was just sixteen miles from Springdale. The school saw its mission as providing "women with an education comparable to that which men could receive at the time at 'colleges of the first class.'" Rachel said she chose PCW because it was a "Christian college founded on ideals of service and honor." PCW also fit the Carsons' limited purse. Her father had planned to help meet school costs by selling lots but was unable to do so. As a result, Maria took on additional piano students and sold produce from the farm. She even sold her family china. Yet, despite winning a competitive state scholarship, Rachel found the annual costs of the college prohibitive until PCW made up the difference. In the fall of 1925, Rachel set off with a debt to both her parents and the college. She had no intention of disappointing either.[19]

PCW offered the intellectual challenges Rachel sought, but all too often it conformed to the social conventions of the day, especially in the first two years. The college served most of its students as a finishing school in which marriage and motherhood, rather than career, lay in the future of its graduates. President Cora Helen Coolidge encouraged her students to dress well and master social skills. To help them prepare, PCW held teas and coed dances on a regular basis, but most of the time Rachel refused to attend. Not only did she lack the proper clothes for many of these events, but she also chose to study in her free time. A severe case of acne that persisted throughout her four years at PCW did not help matters. Still, Carson found her own social outlets. Her roommate, Dorothy Appleby, came from a similar Presbyterian background, and the two often attended Sunday church services together. Rachel played team sports, such as field hockey and basketball, with skill and intensity, despite a slight build that made her less than a formidable opponent.[20]

One other factor limited Carson's social life in a way she could not choose to avoid: her mother's regular Saturday visits. When she might have been socializing with friends and classmates, she and her mother spent time together without including others in their conversations or sharing the cookies Maria baked for the occasion. On the weekends that her mother did not come, Rachel often went home. Homesickness may have been a motive, but more likely, Rachel simply valued her mother's company. Theirs was an exclusive friendship. Maria devoted her visits at PCW to Rachel and to sharing with her the stimulation of a college campus that she had missed as a girl. Those who saw them together never sensed any resentment on Rachel's part. She treated her mother as an intimate friend. On her mother's sixtieth birthday, Rachel was disappointed that they could not be together. She described her college years as "this four year banishment" and looked forward to a future where they could celebrate such occasions together.[21]

For Rachel, PCW brought together the two strands that later defined her career—her ambition to be a writer and her love of nature and science. During her first year, Rachel came under the tutelage of Grace Croff, an assistant professor of English, recently hired from Radcliffe College. Croff was a demanding teacher of freshman composition, and Rachel's promise as a writer caught her attention. By spring their classroom relationship had blossomed into friendship. Croff praised several of Carson's pieces as particularly well-written and encouraged Rachel to join the student newspaper and literary magazine. These compositions combined close observation of nature with the graceful prose she developed in her writing for *St. Nicholas Magazine.* Not unlike Stephen Crane, who wrote intimately about the Civil War without having been a soldier, Carson in "Master of the Ship's Light" evoked the ocean and the seashore, though she herself knew them only through her reading. "Torrents of foam and spray dashed against the sides of Arrowhead Light, whose unwavering beams, like long, white lances, warned approaching ships of the danger," she wrote. Lacking a more visual sense of the sea, she depicted it largely by its sounds and, in a strategy that she employed in her mature writing, turned it into a character: "How fiendishly it shouts, and laughs to itself!"[22]

In her sophomore year, Rachel discovered the scientific half of her intellectual life. Needing to fill a laboratory-science requirement, she enrolled in a biology class. The subject introduced her to a more scientific version of the natural history she learned from her mother. Her biology professor, Mary Scott Skinker, taught her to be a scientist. Students at PCW regarded Skinker with a mixture of awe and admiration. She combined the physical grace and glamour of a movie star with the intellectual rigor of a hard scientist. Rachel fell under her spell, and Skinker was equally drawn to her acolyte. Few students at PCW combined the curiosity, intelligence, and diligence that distinguished Carson. As with Croff a year earlier, this teacher-student relationship turned into friendship. Rachel absorbed every lesson, while Skinker provided a model of the "spiritual adventure" Rachel sought in her own life.[23]

Sometime during her sophomore or junior year at PCW, Carson had what she later remembered as a life-defining insight that connected her love of writing and biology to the sea. One night, as a storm raged outside, Carson was alone reading Tennyson's poem "Locksley Hall." Thunderclaps shook her dormitory as she read a passage that she recalled ever after:

> Cramming all the blast before it, in its breast a thunderbolt
> Let it fall on Locksley Hall, with rain or hail or fire or snow;
> For the mighty wind arises, roaring seaward, and I go.

The last line in particular, "For the mighty wind arises, roaring seaward, and I go," seemed to Rachel to anticipate her calling in life. "That line spoke to something within me," she wrote, "seeming to tell me that my own path led to the sea—which then I had never seen—and that my own destiny was somehow linked to the sea."[24]

No doubt, Tennyson spoke to Carson at a time when she yearned for direction in her life. Yet, she had already developed a taste for sea stories and once told a correspondent "Even as a child—long before I had ever seen it—I used to imagine what [the sea] would look like, and what the surf sounded like." By this time at PCW she was on the verge of making some choices about her future. Her studies had already begun to link the subjects that fed her intellectual curiosity. Her friendships with Croff and Skinker nurtured her self-confidence. So did the small circle of friends she developed among students who shared her intellectual passions. Most PCW students, Carson believed, sought little more than good grades and social acceptance. Only among her faculty mentors and close friends did she discover people she saw as morally superior, those who, like herself, valued the "undaunted efforts of the adventurous mind" over the more pedestrian matters of marriage and social acceptance.[25]

Two somewhat contradictory strands to her personality clashed within her. On the one hand, she was conventional in personal style and literary tastes and would remain that way throughout her life. The bobbed hair, short skirts, and cigarettes that defined the flappers of the "Roaring Twenties" left her cold. She wore the demure blouses and floor-length and full skirts common among proper young women. One of her admirers suggested, "there was something about her of the nineteenth century, of the times when there really had been young ladies. She had dignity, she was serious; and as with Lear's Cordelia, 'her voice was ever soft, gentle and low.'"[26]

Her preference in literature ran to nature writers such as John Burroughs, John Muir, and Dallas Lore Sharp, not "the sordid naturalism which characterizes modern literature." Sharp's stories, some of which appeared in *St. Nicholas Magazine*, Carson wrote in a class essay, "have about them the tang and freshness of a sea breeze, the limpid beauty of a mountain pool. They are, indeed, a part of the 'immortal literature of escape.'" In his work, she believed emotions were "justified by a strong basis of thought; they are vivid, lasting and varied; they tend always to lift the mind up and away from life's sordidness." She applauded his rejection of realism and naturalism: "Art does not watch life and record it. Art loves life and creates it."[27]

Yet, mirroring her nonconforming impulses, these nature writers were rebels in their own way. They rejected the bustle of the urban world and pursuit of material comforts that defined American society in the early twentieth century. Muir liked nothing better than camping under the stars or swaying in the high branches of trees buffeted by windstorms. The eccentric Burroughs kept a fresh woodchuck on hand to serve his visitors at dinner. Theirs was a world infused with the spiritual uplift brought by intimacy with nature. In embracing those writers as literary models, Carson had begun to confirm her character. No matter how conventional in manner and personal appearance, Rachel Carson was, like the writers she admired, something of a subversive.

By the fall of her junior year she had also decided to major in biology. The choice was a conflicted one. Ever mindful of her limited financial resources, Rachel understood that her decision complicated her future and upset the plans she and her mother had laid. Writing had already brought her recognition. Indeed, Maria Carson sent Rachel to PCW to develop that talent. Whatever its intellectual rewards, science offered few career opportunities for women. The best Rachel could hope for was to follow the same career path as Skinker and earn a position teaching at a women's college, a prospect that appealed to her. As their friendship blossomed, Skinker superseded both her mother and Grace Croff as the woman to whom Rachel most turned for guidance and friendship. By the end of her junior year, Carson no longer agonized over her choice for biology. "I don't know why I ever hesitated about it," she wrote her friend and fellow biology major Mary Frye. "Did I tell you that I'm hoping to go right on and get my master's degree?"[28]

Rachel's enthusiasm left her unprepared for the shock she felt, when she learned that Skinker was taking a leave to complete her doctorate. That meant she would not return to guide Rachel through her final year. Carson suspected that the college president had pressured Skinker out of PCW. The two disagreed on matters ranging from standards to curriculum to the role of women in science. Skinker also decided that graduate work took precedence over marriage and ended a relationship with a suitor. Nothing would help Skinker punctuate the break better than leaving town. Despite her dismay, Carson believed that, in forsaking PCW and marriage, her mentor was making the right choices.

Rather than wallow in self-pity, Rachel decided to skip her senior year and follow Skinker to Johns Hopkins. One insuperable problem frustrated Rachel's plan; she could not afford to transfer. Hopkins did admit her to its biology graduate program, though without the financial support she needed to leave PCW. Worse yet, Grace Croff also left the college. Rachel never learned why, but she blamed the college nonetheless. When she graduated from PCW in 1929 with honors, she was grateful for her education and the financial support the college provided her but resentful that it drove away the two people she most admired. Carson never forgave her alma mater for what she saw as a betrayal of her mentors and the values she shared with them.[29]

After she left PCW, Mary Scott Skinker continued to guide Rachel's early career. The two women remained friends for life. Skinker spent the summer of 1928 at the Marine Biology Laboratory (MBL) in Woods Hole, Massachusetts. Her letters to Rachel glowed with the enthusiasm of a scientist immersed in her work. Rachel told her friend Mary Frye that Woods Hole "must be a biologist's paradise" and decided that she and Mary would somehow find a way to work there after graduation.

During her senior year, Carson reapplied to Johns Hopkins. The graduate school admitted her to its zoology program with full tuition and a $200 annual stipend. In choosing science, Carson believed she had forsaken her career as a writer. One of her friends was less certain. Marjorie Stevenson urged her "not to take all the frogs and skeletons too seriously." "Remember," Stevenson wrote in her yearbook, "I prophesy you'll be a famous author yet."[30]

That summer Carson realized her ambition to study at Woods Hole. The MBL was everything she had hoped for. Among the seventy-one "beginning investigators" at the laboratory were thirty-one women, including Rachel and Mary Frye. Rachel's project involved a study of the cranial nerves in reptiles that she hoped would become the basis for her research at Johns Hopkins. She was dismayed to discover that the training she received during her last year at PCW left her unprepared for the research program at MBL. But whatever insecurities she felt were more than compensated for by the social and intellectual life there. Female scientists mingled freely with their male counterparts, and graduate students worked alongside eminent profes-

*Rachel first experienced the sea during her first summer at the oceanographic lab at Woods Hole in Massachusetts. (*SOURCE: *Courtesy of the Rachel Carson Council.)*

sors. Among them was Rheinart Cowles, a marine biologist from Johns Hopkins and her future graduate advisor. Beach parties, picnics, and family-style dinners offered opportunities to partake in freewheeling intellectual conversations.

Above all, the summer spent at Woods Hole connected Rachel to the sea about which she had so often dreamed as a girl. Walks in the light of the full moon along the shoreline at low tide had a magical quality for her as she observed the creatures clinging to the rocks and swaying in the tide pools. She later wrote of that summer,

> There [MBL], too, I began to get my first real understanding of the real sea world—that is, the world as it is known by shore birds and fishes and beach crabs and all the other creatures that live in the sea and along its edges. At Woods Hole we used to go out in a little dredging boat and steam up and down the Vineyard Sound or Buzzards Bay. After a time, with much violent rocking of the little boat, the dredge would pull up and its load of sea animals, rocks, shells, and seaweed spilled out onto the deck. Most of the animals I had never seen before; some I had never heard of. But there they were before me. . . . Probably that was when I first began to let my imagination go down under the water and piece together bits of scientific fact until I could see the whole life of those creatures as they lived them in that strange sea world.[31]

The graduate school at Johns Hopkins was a different world altogether. It stressed pure research, and students in Carson's position were expected eventually to become doctoral candidates. There were fewer instances of gender discrimination there than at other science research centers and more women, but the press of work left Rachel little time for socializing with women or men. She made few friends among her fellow researchers. Instead, she made the more consequential decision to reunite her family.

In the fall of 1929, shortly after she enrolled at Hopkins, the stock market crashed, leaving her family's financial situation ever more uncertain. Rachel decided that circumstances were more promising in Baltimore than in Springdale. Besides, she and her mother had been separated for some nine months. Prior to the summer of 1929, they had never been farther apart than the sixteen miles that separated Springdale from the PCW campus. On top of that, housing her parents, brother Robert, and sister Marian and her two daughters under one roof would save money. For some reason, the potential emotional baggage and family complications did not concern her.

The first house she picked shared some of the qualities of the Springdale farm, with the added advantage that it was larger and equipped with indoor plumbing. What most appealed to Rachel were the surrounding woods and the proximity to Chesapeake Bay, just two miles away. The only disadvantage was the thirteen-mile commute to the Hopkins campus. More important to Rachel, her family's arrival in Baltimore restored her companionship with her mother. She thrived on the support Maria provided—much of it practical. Rachel never had much appetite for housework and cooking. While she could best meet the family's financial needs, Maria could manage her household. Thus, in 1930, they settled into the arrangement that they maintained until Maria's death. Rachel was the breadwinner and writer and Maria, the housekeeper and secretary. The composition of the household and the address changed from time to time, but the roles and the family living arrangements remained the same.[32]

The Depression era was not the time for Rachel or any other woman to launch a career, especially in the world of science, where men ruled the laboratories and academic departments. Nor did she show unusual promise as a researcher. At Hopkins she proved herself once again to be a diligent student who excelled in course work. But Hopkins was a research center, and Carson quickly discovered that "the lab is my world and is going to be my chief existence until I get my degree." Given the growing financial burden imposed by her family, the sooner she earned that degree, the better. A tuition increase for the 1930–1931 academic year compounded her problem. A summer teaching stipend barely covered the cost of rent and debt payments to PCW. She now had to work part-time in order to continue her studies. When her research work failed to produce substantial results, her degree was delayed another year. At that point, she shifted from comparative evolution to marine biology, a field into which her advisor, Professor Rheinart Cowles, had recently shifted.[33]

Carson's training as a scientist flowered in her master's thesis, "The Development of the Pronephros During the Embryonic and Early Larval Life of the Catfish (*Ictalurus punctatus*)." But her gifts as a nature writer were nowhere to be found. The opening sentence groaned under the weight of academic prose: "Anatomists and embryologists are almost universally agreed that the excretory function in adult Teleosts is performed by organs of higher specialization than the pronephros." Cowles praised her review of the scientific literature on her subject and the rigor of her laboratory findings. Nonetheless, some senior members of the zoology faculty doubted that she had the makings of a research scholar. Herbert Spencer Jennings, the department's eminent geneticist, who prized research above all, damned her with the faintest praise. Carson, he wrote in a recommendation letter, was "a thorough, hard-working person, not brilliant, but very capable and with a good knowledge of biology." He concluded that she would "continue to be a satisfactory teacher."[34]

Even if Carson hoped to reverse that judgment, circumstances conspired against her ambition to complete her doctorate. In 1932, after she had earned her master's degree, her family burdens increased. The possibility that brother Robert would contribute to the maintenance of the household ended when he moved out on his own. Diabetes left her sister Marian disabled, and her father was growing frail. In 1934, Rachel finally dropped out of the doctoral program and began to look for a full-time teaching position. Writing offered some promise of additional income, but that hope initially produced no immediate results. Reworking her college poems and short stories, she sent them to leading magazines, including *Reader's Digest, The Saturday Evening Post*, and *Collier's*. Back came an impressive list of rejection slips.

Then tragedy struck. On a July day in 1935 her father complained of feeling unwell and stepped out into the yard for a breath of air. Moments later he fell dead, the victim of a massive heart attack. The family's finances were so strained no one could afford to attend his funeral. Maria shipped his body to Pennsylvania, where his three sisters arranged the burial. He had never been a defining presence in Rachel's life, but theirs was a close family, bound by ties of affection; she could not help but be shaken by his loss.

At this desperate moment, Mary Scott Skinker once again set Carson on a new path. Skinker had made a comfortable place for herself as a research zoologist in a government agency in Washington, D.C. She persuaded Carson to take civil service exams in wildlife, aquatic biology, and parasitology. Carson scored well on all three. Skinker then urged her to renew contact with Elmer Higgins, an official in the U.S. Bureau of Fisheries, whom she had met several years earlier. Here, as it turned out, was a case where "luck favored the prepared mind."

Higgins was responsible for a bureau-sponsored radio series his colleagues dismissively referred to as "seven-minute fish tales." The series became a headache for Higgins. No one in the bureau knew how to make marine biology interesting to a general radio audience. Though he had no regular job for Carson, he asked her to write a few scripts. Eight months later, she completed the series to the great satisfaction of Higgins and other bureau officials. Higgins then had her write an introduction to a government brochure on marine life. When he met her in April 1936 to discuss the piece, his reaction stunned her. The essay would not do, he said. "The World of Waters" was too good for a government brochure. She should write a new introduction, he advised, "but send this one to the *Atlantic*."[35]

Carson later expressed her gratitude to Higgins, acknowledging him as "really my first literary agent." A year passed, however, before she acted on his suggestion. For one thing, Higgins found her a full-time position. Her high scores on the civil service exams qualified her as a junior aquatic biologist in the Bureau of Fisheries Division of Scientific Inquiry. Robert Nesbit, who became her immediate boss, headed a team collecting data on the fish populations of Chesapeake Bay. Carson analyzed the data they produced, wrote their reports, and created brochures that informed the public about their findings. In the course of her work, she visited laboratories and field stations and, in the process, began to form a network of contacts among researchers in marine biology. As an outlet for her literary talents, she sold articles on marine life to the *Baltimore Sun*. The mundane topics included shad populations, oyster beds, and the fishing industry; but her theme was often the same—marine ecologies in some state of crisis. The *Sun* editor so admired her ability to present scientific information in ways that general readers could appreciate that he paid her $20 for each story he published.[36]

Life was good, but only for the moment. Carson had work that took her into the field often enough to expand her horizons, and the *Baltimore Sun* was enthusiastic about most of her ideas for stories. Yet, in each phase of her life a dark shadow seemed to follow her. At PCW she had hitched her star to Mary Scott Skinker, only to have Skinker leave the college. Financial pressures terminated her graduate work. Now the problem was her sister Marian. As a diabetic, Marian had difficulty holding even a part-time job and needed help raising her two young daughters, Virginia and Marjorie. Whether from the disappointments of two failed marriages, poor health, or financial woes, Marian suffered recurring bouts of depression. Early in 1937 she came down with pneumonia. Within weeks she was dead, having just passed her fortieth birthday. Rachel, not yet thirty, and Maria, already over seventy, decided to raise the two girls. They had little choice. Robert had turned his back on the family, and Marian's husband had abandoned his children. What else could these women do?[37]

Rachel felt the weight of this new responsibility. Despite her love for the woods and Chesapeake Bay, she moved her family into a new house in Silver Spring, Maryland. Though more expensive, it was closer to her office in nearby College Park. One way to offset new financial pressures was to earn more from her writing. Carson turned once again to the story that for almost a year had been tucked away in her desk drawer. Always a perfectionist in her writing, she revised it one more time before sending it off to the *Atlantic Monthly* in early June. A month later, acting *Atlantic* editor Edward Weeks replied that he thought "The World of Waters" was "uncommonly eloquent" and predicted that it would "fire the imagination of the layman." The *Atlantic* wanted to publish the piece in the summer months, when interest in the oceans peaked. Weeks asked only for several small revisions and a new title—"Undersea." He offered $100, and a delighted Carson gratefully accepted.[38]

In publishing "Undersea" in September 1937, the *Atlantic* introduced one of the major literary voices of the twentieth century. In just four pages, Carson demonstrated her ability to appeal to both the poetic imagination and scientific curiosity. "Who has known the ocean," she asked her readers with uncommon eloquence.

Neither you nor I, with our earthbound senses, know the foam and surge of the tide that beats over the crab hiding under the seaweed of his tide-pool home; or the lilt of the long, slow swells of mid-ocean, where wandering shoals of fish prey and are preyed upon, and the dolphin breaks the waves to breathe the upper atmosphere. Nor can we know the vicissitudes of life on the ocean floor, where the sunlight, filtering through a hundred feet of water, makes but fleeting bluish twilight, in which dwell sponge and mollusk and starfish and coral, where swarms of diminutive fish twinkle through the dusk like a silver rain of meteors, and eels lie in wait among the rocks. Even less is it given to man to descend those six incomprehensible miles into the recesses of the abyss, where reign utter silence and unvarying cold and eternal night.

The sea world differed from the terrestrial domain of humans. The ocean's creatures, Carson wrote, live in a fluid world: "It is water that they breathe; water that brings them food; water through which they see, by filtered sunshine from which the first red rays, then the greens, and finally the purples have been strained; water through which they sense the vibrations equivalent to sound." She asked her readers to consider the paradoxes that define this wondrous place, home to both the "great white shark, two-thousand-pound killer of the seas, and the hundred-foot blue whale, the largest animal that ever lived." Yet, "it is also the home of living things so small that your two hands might scoop up as many as there are stars in the Milky Way." More marvelous still, all creatures in the sea, from the largest to the smallest, depend on microscopic plankton for survival. Here mingle the elements "which in their long and ancient history, have lent life and strength and beauty to a bewildering variety of living creatures."[39]

To Carson, this mingling was not happenstance but part of a larger plan. The seas store the materials on which life, energized by the sun, begins anew each spring. Hungry plankton that feed on the microscopic plants are in turn sustenance for the fish: "all in the end to be redissolved into their component substances when the inexorable laws of the sea demand it. Individual elements are lost to view, only to reappear again and again in different incarnations of material immortality." Individual creatures may die, but their bodies sustain life "in a panorama of endless change." Here, though Carson did not describe it as such, ecology was at work. From contemporary ecologists she had also learned about "the food chain," first introduced by English zoologist Charles Elton in his 1927 book *Animal Ecology*. In every community on land or sea, plants use photosynthesis to turn sunlight into food. Those at the bottom of the chain feed those further up. "Each stage in an ordinary food-chain has the effect of making a smaller food into a larger one, and so making it available to a larger one," Elton observed. Eventually, even the most complex organisms decompose into the matter that sustains future plant life. Thus, without ever resorting to scientific labels or technical jargon, Carson exposed her readers to elements of what would soon be known as "the new ecology."[40]

Scientists were not the only ones who appreciated Carson's literary sleight of hand. Among the admiring readers of "Undersea" were Quincy Howe, an editor at the Simon and Schuster publishing house, and Hendrik Willem Van Loon, the author of a major

1920s bestseller, *The Story of Mankind*. Together, the two persuaded Carson to turn "Undersea" into a full-length book, though in truth, she needed little convincing. Necessity now sharpened the writer's ambition that Maria Carson had nurtured in Rachel as a girl. The attention of two such eminent literary figures was all the encouragement she needed.

Her book, Carson explained to Van Loon, would avoid the anthropocentric or "human bias" that infected most writers about the sea. "The fish and other sea creatures must be central characters and their world must be portrayed as it looks and feels to them—and the narrator must not come into the story or appear to express an opinion. Nor must any other humans come into it except from the fishes' viewpoint as a predator and a destroyer." This would be the story of the sea and its creatures. "The ocean is too big and vast and its forces too mighty to be much affected by human activity," she believed. Time proved her wrong on this last observation, as overfishing, pollution, and other by-products of human society seriously altered the ocean's ecology. But even at this point she began to suspect that naturally occurring chemicals, such as fluorides and selenium, as well as coastal pollution were harmful to aquatic life.[41]

Deciding to write a book was one thing; finishing it was quite another. Carson frequently admitted that she was a slow writer under the best of circumstances. With work and family obligations competing for her attention, time was barely manageable. Promotions at the Bureau of Fisheries increased both her pay and her responsibilities. So did the reorganization of her agency. Secretary of the Interior Harold Ickes, one of Washington's fiercest bureaucratic imperialists, pulled the Bureau of Fisheries from the Commerce Department to join it with the Agriculture Department's U.S. Biological Survey. In 1939, he brought them into the Interior Department as the U.S. Fish and Wildlife Service. Ickes saw this as one step toward his larger plan of turning Interior Department into a Department of Conservation. For Carson, the added work of the new agency left time to write only on weekends and evenings.

Still, work had its compensations. The newly reorganized Fish and Wildlife Service operated research facilities up and down the Atlantic coast. For several summers she was able to return to Woods Hole and expand the material for her book. In July 1938, she took her entire family to the research station at Beaufort, North Carolina, along the Outer Banks. The remote beaches and rolling dunes inspired her. Day and night, notebook in hand, she studied the seascape that would become the background for the shorebird section of her new book. She jotted down the physical characteristics and the sounds of creatures she hoped to write about. She had time to reflect on what she observed. One note spoke of the "urgency of [the birds'] spring migration." This drive to reproduce became a central theme in her book. "Buzzard on beach near porpoise—attracted by scent—do they fly over to see if shadow startles animal?" she wondered in another entry. A more imponderable question sometimes crossed her mind: "How does breaking wave (a) sound and (b) look to sanderlings?" The sheer magnitude of the ocean's fertility amazed her, as when she observed the "shoreward migration of mackerel school one mile wide—20 miles long." The sights and sounds of the sea evoked the poet in her: "The crests of the waves just before they toppled caught the gold of the setting sun then dissolved in a mist of silver. The sand in the path of each receding wave was amethyst, topaz, and blue-black."[42]

Not until the spring of 1940 was Carson able to finish enough of the book to secure a contract from Simon and Schuster. Facing a December deadline, she began to write in earnest. Fortunately, she had moved the family to another house in Silver Spring that contained a single, large upstairs space she could use as both office and bedroom. The months when she wrote *Under the Sea-Wind* stuck in her memory as a period of intense creativity. For this she had her mother to thank. Maria managed the household so efficiently that Rachel worked without interruption. Only her two cats, Buzzie and Kito, were allowed to violate this splendid isolation. When a section of writing was finished or needed revising, Rachel read it out loud to her mother. Then, while Rachel was at work and the girls in school, Maria retyped the copy so it was on Rachel's desk when she returned at night. Most sections went through multiple revisions and endless fine-tuning. But her determination and disciplined approach paid off. By early November she mailed a completed draft to Simon and Schuster.[43]

One year later, on November 1, 1941, *Under the Sea-Wind* appeared in print. Appropriately, Carson dedicated the book to Maria and gave a copy to Elmer Higgins, with the inscription "To Mr. Higgins, who started it all." Anxious about potential sales, she wrote to Van Loon seeking reassurance. With his usual whimsy, he reminded her that it was hard to guess "what the public will swallow or not" but added that he hoped "this time they prove to be fond of fish."

Certainly, most of the reviewers turned out to be exceptionally "fond of fish." They suggested that the book revealed an unusual talent and stressed that it was both well informed and beautifully written. A reviewer in the respected *Christian Science Monitor* observed, "An ocean and its shores will never again seem deserted or lonely to one who reads *Under the Sea-Wind*, for through the pages of this enthralling and unusual book one is made aware of the myriads of interacting creatures both beneath and beyond the surging waves." Carson's gender also drew attention, with frequent reference to her as a "woman naturalist" or "woman biologist." Yet, gender did not prevent Carson from exposing the violence that characterized life in what to most human observers were tranquil seas. A *Scientific Book Club Review* noted, "There is ruthlessness as well as beauty in nature. . . ." Perhaps the most touching and sensitive response came from Mary Scott Skinker, who wrote "I can think of nothing in years that I've wanted as much to have this come to you." Skinker understood too that Maria was basking in Rachel's success. "Much of my joy finds its source in thinking of your mother's happiness."[44]

Carson's decision to tell the story from the creatures' point of view and to make the sea itself a central character proved an inspired one. As readers followed the life cycles of seabirds, mackerels, and eels, they found themselves exploring a world long invisible to the human eye. The sanderlings Carson observed on North Carolina's Outer Banks spent the better part of their lives in the Canadian Arctic and the windswept plateau of Patagonia in southern Argentina. Twice a year some mysterious force compelled them to venture on a 12,000-mile journey. Along their way they encountered all manner of dangers that Carson described in gripping detail. She told the story of the sanderling Blackfoot who suddenly heard "Tee-ar-r-r! Tee-ar-r-r!"—the cry of an attacking tern: "The swoop of the white-winged bird, which was twice as large as the sanderling, took Blackfoot by surprise, for his senses had been occupied with avoiding the onrush of water and preventing the escape of the large crab he held in his bill." Blackfoot sprang into the air over the water with the tern in pursuit:

> In his ability to bank and pivot in the air Blackfoot was fully the equal of the tern. The two birds, darting and twisting and turning, coming up sharply together and falling away again into the wave troughs, passed out beyond the breakers and the sound of their voices was lost to the sanderling flock on the beach.

In a theme that Carson developed throughout the book, fate as much as instinct or skill sealed the outcome. As the tern closed on Blackfoot, it spied "a glint of silver in the water" and turned its flight into a sharp descent from which it emerged "with a fish curling in his bill." Blackfoot, having been forgotten, returned to the sanderlings feeding along the beach.

Maria Carson's lessons in close observation shaped this writing. From Blackfoot's anatomy to his feeding habits, defense mechanisms, and vocal patterns, Carson had the details just right. She also knew how to construct an engaging plot. Blackfoot's sojourn on the Outer Banks was but a brief chapter in the odyssey on which he and the other sanderlings had embarked. Having fattened themselves on the rich offerings of the Carolina shore, they set off once more toward the Arctic. There, they encountered new predators, a harsh terrain, and the unexpected resurgence of winter's bitter weather. In the rituals of his species, Blackfoot found a mate, Silverbar, who hatched four chicks. But as Carson made clear to her readers, danger lurked everywhere.

Following the instincts of her species, Silverbar carried off the shells of her newly hatched chicks. "So countless generations of sanderlings had done before her," Carson noted, "by their cunning outwitting the ravens and foxes." Neither the predatory falcons nor jaegers saw "the little brown-mottled bird as she worked her way, with infinite stealth, among the clumps of betony or pressed her body closely to the wiry tundra grass."

Even with all that effort, the nest was no longer safe. Foxes could catch the faint scent of the chicks. So for two weeks Silverbar herded her brood about until they were ready to fly. At that point, the time had come to teach them to survive on their own. Sometime in August, the cycle began anew as Silverbar and Blackfoot prepared to leave the Arctic:

> The nesting was done; the eggs had been faithfully brooded, the young had been taught to find food, to hide from enemies, to know the rules of the game of life and death. Later, when they were strong enough for the journey down the coast lines of two continents, the young birds would follow, finding the way by inherited memory. Meanwhile, the older sanderlings felt the call of the warm south; they would follow the sun.

Along the way the sanderlings joined a wide array of other birds heading toward their winter nesting grounds. For Carson this story demonstrated how in the cycles of these creatures nature preserved its balance.[45]

Despite her focus on sea creatures, Carson was neither indifferent to people nor hostile to them. She understood that human beings had their own roles to play in the cycles of nature. Given the fecundity of the sea, those roles were not necessarily destructive. Humans were but one more predator with whom the creatures of the sea contended. To a stroller on a Carolina beach the north winds of fall meant stinging sand blown into eyes and hair. To the commercial fishermen the same wind meant fish for their nets. When it blew, they sprang into action. As the tide turned, a boat shot out from the beach to set the nets. The target was mullets, thousands of slim, silver-gray fish that inhabited the coastal waters. The net formed into a semicircle with a rope on each end tight in the hands of the men on the shore. Carson described the primal struggle that ensued:

> Milling in a frantic effort to escape, the mullet drive with all their combined strength of thousands of pounds against the seaward arc of the net. Their weight and the outward thrust of their bodies lift the net clear of the bottom, and the mullet scrape bellies on the sand as they slip under the net and race into deep water. The fishermen, sensitive to every movement of the net, feel the lift and know they are losing fish. They strain the harder, till muscles crack and backs ache. Half a dozen men plunge into water chin-deep, fighting the surf to tread the lead line and hold the net to the bottom.

Eventually, the fishermen got the upper hand. The net formed into "a huge, elongated bag, bulging with fish." As they pulled it onto the beach, "the air crackles with a sound like the clapping of hands as a thousand head of mullet, with the fury of their last strength, flap on the wet sand."

Haste made waste, Carson noticed. Among the mullet were other fish—sea trout, pompano, young mullets, ceros, sheepshead, and sea bass—"too small to sell, too small to eat." Rushing to store their catch, the fishermen threw these unwanted species onto the beach. There, amidst the litter, some returned on the waves to the ocean, while most died stranded beyond the water's reach. "Thus, the sea unfailingly provides for the hunters of the tide lines," she concluded. What was waste for the fishermen was food for the shore creatures. First came the gulls, followed by the ghost crabs and the sand hoppers. In time, they would reclaim "to life in their own beings the materials of the fishes' bodies." To Carson this process was all part of nature's plan: "For in the sea nothing is lost. One dies, another lives, as the precious elements of life are passed on and on in endless chains." Not even the fishermen disturbed that plan. The ocean's bounty was such that no species threatened the survival of others. While the fishermen gathered around their stoves to fight off the night chill, "mullet were passing unmolested through the inlet and running westward and southward along the coast. . . ."[46]

In *Under the Sea-Wind,* Carson preached a message of optimism: while individual life might be finite, life in nature cycled in enduring rhythms. That lesson, she told an interviewer, could apply to humans as well: "Each of these stories seems to me not only to challenge the imagination, but also to give us a little better perspective on human problems. They [stories] are as ageless as sun and rain, or the sea itself. The relentless struggle for survival in the sea epitomizes the struggle of all earthly life, human and non-human."[47]

NOTES

1. Hendrik Van Loon to Carson, September 10, 1937, RCP–BLYU.
2. Liner notes, *Under the Sea-Wind;* see also review files RCP–BLYU.

3. Brooks, *House of Life*, p. 18; Carson to Curtis Bok, July 12, 1956, RCP–BLYU.

4. The best source for Carson's family and early childhood is Lear, *Rachel Carson*, pp. 7–18.

5. Some biographic materials are available in Carson's papers. See, for example, Maria Carson's brief account of Rachel's first months; Carson left a biographical fragment from around 1954; and she provided some other materials in "The World Around Us," a speech to Theta Sigma Phi, April 21, 1954—all available at RCP–BLYU.

6. Speech to Theta Sigma Phi; see also Brooks, *House of Life*, p. 18.

7. Lear, *Rachel Carson*, p. 7.

8. Carson, "The Little Book for R. W. Carson," RCP–BLYU.

9. Carson, "A Battle in the Clouds," *St. Nicholas Magazine*, Sept. 1918, 45, p. 1048.

10. Carson to Author's Press, July 13, 1921, RCP–BLYU.

11. Worster, *Nature's Economy*, p. 7. Worster provides a framework in which to see how Carson's thinking about nature and science fits into a wider context of ecological thought.

12. Ibid., pp. 7–8.

13. Peter Schmitt, *Back to Nature: The Arcadian Myth in Urban America* (New York, Oxford University Press, 1969), p. 87. The link between nature studies and reformed Protestantism is developed in Stoll, *Protestantism, Capitalism, and Nature in America*.

14. Schmitt, *Back to Nature*, pp. 30, 96.

15. Worster, *Nature's Economy*, p. 185; Lear, *Rachel Carson*, pp. 14–15.

16. Carson, "My Favorite Recreation," *St. Nicholas Magazine*, July 1922, 49, p. 999.

17. Parnassus High School *Yearbook*, 1925; see also Brooks, *House of Life*, pp. 18–19, and Lear, *Rachel Carson*, pp. 23–24.

18. Carson, "Intellectual Dissipation," senior thesis, June 1925, RCP-BLYU.

19. The information on Chatham College came from the college website; see also Carson, "Who I Am and Why I Came to PCW," 1925, RCP–BLYU, and Lear, *Rachel Carson*, pp. 25–26.

20. Lear, *Rachel Carson*, pp. 27–30.

21. Carson to Maria Carson, February 1929, RCP–BLYU; see also Lear, *Rachel Carson*, p. 31.

22. Carson, "The Master of the Ship's Light," May 26, 1926, RCP–BLYU.

23. Lear, *Rachel Carson*, pp. 35–38.

24. Carson, Autobiographical piece, RCP–BLYU; see also Lear, *Rachel Carson*, p. 40, and Brooks, *House of Life*, pp. 20–21.

25. Carson, "They Call It Education," 1928, RCP–BLYU. For an interesting rereading of this episode, see Howarth, "Turning the Tide," p. 44.

26. Brooks, *House of Life*, p. 98.

27. Carson, "Dallas Lore Sharp," May 28, 1926, RCP–BLYU.

28. Lear, *Rachel Carson*, p. 45.

29. Ibid., pp. 46–47.

30. Ibid., p. 5.

31. Carson to Mrs. Eales, undated memo on *Under the Sea-Wind*, c. 1941, Box 3, RCP–BLYU.

32. Lear, *Rachel Carson*, pp. 66–67.

33. Ibid.

34. Carson, "The Development of the Pronephros During the Embryonic and Early Larval Life of the Catfish (*Ictalurus punctatus*)," June 1932, RCP–BLYU; see also Lear, *Rachel Carson*, p. 76.

35. Brooks, *House of Life*, pp. 22–23.

36. Carson's correspondence with the *Baltimore Sun* is collected in her papers, RCP–BLYU.

37. Brooks, *House of Life*, p. 22; Lear, *Rachel Carson*, pp. 84–85.

38. Edward Weeks to Carson, July 13, 1937, RCP–BLYU.

39. Carson, "Undersea," pp. 322–325.

40. On Elton and the new ecology, see Worster, *Nature's Economy*, pp. 294–315. Carson later acknowledged her familiarity with Elton in a letter to Harvard entomologist and social biologist E. O. Wilson. I am most indebted to Frederick Davis of Florida State University who established this connection. Frederick R. Davis, "'Like a Keen North Wind': How Charles Elton influenced *Silent Spring*," delivered at the American Society for Environmental History, St. Paul, Minnesota, April 1, 2006. See Carson to Wilson, November 6, 1958, RCP–BLYU.

41. Carson–Van Loon correspondence, September–October 1937, RCP–BLYU.

42. Carson, Preliminary notes 1938, Box 1, Folder 1, RCP–BLYU.

43. Carson to Curtis Bok, RCP–BLYU; Lear, *Rachel Carson*, p. 102.

44. The reviews are collected in Carson's papers, Box 3, and from Mary Scott Skinker to Carson, September 23, 1941, RCP–BLYU.

45. Carson, *Under the Sea-Wind*, pp. 19–41.

46. Ibid., pp. 53–54.

47. Carson memo to Eales, c. 1941, RCP–BLYU.

· Two ·

SUMMER

Florescence: *The Sea Around Us*

FOR RACHEL CARSON, 1941 SHOULD HAVE ENDED IN TRIUMPH. *Under the Sea-Wind* came out in November to glowing reviews from literary and scientific readers. In the book Carson married her dream of being a writer with her love of the sea. But circumstances once again conspired to rob her of success. This time the source was not personal but a world torn by war. On December 7, one month after the book was published, the Japanese bombed Pearl Harbor. As war fever swept the nation, Mary Scott Skinker captured the feelings of many Americans when she wrote to Carson two days later, "THE WORLD tonight IS DEPRESSING, and thoughts of friends in danger serve but to increase a sense of despair over the inevitable period of years we must face before we know anything resembling peace and security."[1]

Within six months Carson learned how much the coming of World War II distracted the reading public from her story of the oceans and their creatures. "The world received [the book] with superb indifference," she later remarked. As of June 1942, sales stalled at barely over 1,200. Somewhat ruefully, she noted that the "rush to the bookstore that is every author's dream, never materialized." Although she offered her editor ideas, Simon and Schuster did nothing to promote the book. Wartime shortages, including paper, quashed prospects for a British edition. She urged Simon and Schuster to nominate it for literary awards that carried cash prizes. Her hopes for a Pulitzer Prize were dashed when she realized *Under the Sea-Wind* would be nominated in the category of a "history of the Atlantic shore," a topic that had little to do with her book's study of natural history. By the time Simon and Schuster remaindered the book in 1946, she received less than $700 in royalties. Recognizing her commercial failure, she satisfied herself with the praise of critics, friends, and colleagues. To her publisher she acknowledged, "I have been so pleased with the reception the purely scientific people, who so often have little patience with popularizations of science, have given the book." The commendations of one normally caustic fish biologist "meant a good deal to me," she confessed.[2]

Carson never succumbed to disappointment. Toughness of character, resilience, and an appetite for new experiences sustained her through the book's lackluster sales. So did practical considerations. There was a war to be won, and there were family responsibilities to be met. And she still worked at the Fish and Wildlife Service, which, like all government agencies, mobilized for the war effort, though it was hardly in the line of fire. A plan emerged in 1942 to move the service's operations to Chicago so that more essential agencies could use its office space in Washington, D.C.

"I am quite distressed about it," Carson wrote a friend, "mostly for personal reasons like leaving behind certain people, but there seems to be nothing to do but go along." Fortunately for Carson, the move lasted only six months before the service returned to Washington.[3]

Serious reservations about her job, rather than its temporary uncertainty, prompted Carson's discouragement with her work. For one thing, she wanted to make more of a contribution to the war effort. She told a friend, "I'd rather get into some sort of work with more immediate value in relation to the war, but in boondoggling Washington, I don't know how to find it." Her immediate responsibility was the publication of a series of pamphlets introducing American housewives to new sources of fish protein, scarcely the stuff of a red-blooded warrior. Still, the substitution of these species would add to the nation's food supply. In a line that demonstrated her belief that an informed citizenry made wiser choices, Carson explained, "Our enjoyment of these foods is heightened if we also know something about the creatures from which they are derived. . . ."[4]

Whatever reservations she harbored about her job with the Fish and Wildlife Service, Carson steadily moved up through the agency's ranks. In just three years she went from assistant to associate and then to full aquatic biologist. Each promotion brought a small raise in salary along with a substantial increase in administrative responsibilities. Success did not, however, lead her into the laboratory or to field research for the agency. Those roles were largely closed to women. Instead, she supported the work of her male colleagues by editing their reports and getting them into print. Carson and her six assistants managed all the publishing for the service's information program. She compared it to "the work of a small publishing house." Her success reflected qualities evident in her writing. She was thorough, clear, and sometimes eloquent, even when discussing mundane topics. From her assistants she demanded the best. In turn, they found her fair, if firm, and always calm and courteous. She could also be fun. Shirley Briggs, a junior colleague and lifelong friend, recalled that Carson "could instill a sense of adventure into the editorial routine of a government department." Carson and several friends on the staff often met for lunch in her office. Huddling around an illicit hot plate, they turned "intransigent official ways, small stupidities, and inept pronouncements" into sources of mirth.[5]

In her dealings with John Ady, who guarded the agency's budget, a tougher side emerged. Final approval for all departmental printing rested with Ady. Where Carson sought to improve the look of Fish and Wildlife publications, Ady resisted almost any suggestion that increased printing costs. Each new series meant renewed conflict, as Carson set off to "do battle with Ady." Despite his parsimony, she won occasional concessions because, as Briggs observed, she could be "an unmovable rock" when necessary and knew "when to push and when to wait."[6]

As administrative duties demanded more of Carson's time and energy, she found little opportunity to write. The few articles she did publish were based on military research materials she edited for Fish and Wildlife. One notable piece, "The Bat Knew It First," provided what the Department of the Navy called "one of the clearest explanations of radar yet made available to the public." Such minor success did not ease Carson's mounting frustration. She longed to leave her job to find "something that will give me more time for my own writing." If "I could choose what seems to me the ideal existence," Carson confessed, "it would be to live by writing." Always the pragmatist and given to rigorous self-criticism, she felt she had "done far too little to dare risk it." Providing for her nieces and her mother took precedence over personal ambitions.[7]

Nonetheless, Carson made a final push to find a job that might promote her ambition to be a full-time writer. Through her publisher, she got in touch with DeWitt Wallace, owner of the widely circulated *Reader's Digest*. Carson presented herself as "that comparatively rare phenomenon, a scientist who is also a writer." Besides furnishing her with experience in judging manuscripts, her science training and government work taught her to understand technical language without using it in her writing. In that way she could reach the general public. Wallace found her qualifications impressive but told her that the *Digest* had no openings.[8]

In October 1945 Carson contacted William Beebe, an oceanographer and ornithologist who directed tropical research at the New York Zoological Society. Beebe gained fame as the designer of the bathyscaphe, a small submarine used in deep-ocean research. He had been an early supporter of Carson's work and included two chapters from *Under the Sea-Wind* in an anthology he published of America's great naturalist writers. Beebe also provided her with entrée to other leading figures in oceanography. Carson asked him about the intention of the society to expand its public education program in natural history and the "conservation of living resources." "I don't want my thinking in regard to 'living natural history' to become set in molds which necessity sometimes imposes on Government conservationists," she told him. And giving a hint at her growing frustration, she confessed, "I cannot write about these things unless I can be sincere."[9]

None of these initiatives freed Carson from her job. The positions that did exist generally went to returning veterans and, in any case, publishing, like science, was a field dominated by men. Instead, she satisfied herself with occasional pieces for the magazine market rather than with the books she longed to write. Ironically, short-term frustration may have nurtured long-term success. Despite its demands, the Fish and Wildlife Service began to widen her horizons. It gave her access to cutting-edge research in oceanography that came during the war. Amphibious landings and submarine warfare required careful study of tides, marine life, subsurface geography, and ocean currents. Jungle warfare increased the demand for new means of germ and pest control. At one point, she wrote *Reader's Digest* about research going on "practically at my backdoor" at Patuxent Research Refuge in nearby Laurel, Maryland. "It's something that really does affect everybody." That something was the new miracle pesticide, DDT. "We have all heard a lot about what DDT will soon do for us by wiping out insect pests," she told a *Digest* editor in July of 1945. Researchers at Patuxent wanted to determine "what it will do to insects that are beneficial or even essential; how it may affect waterfowl, or birds that depend on insect food; whether it may upset the whole delicate balance of nature if unwisely used."[10]

Without expressing her skepticism, Carson sensed that the claims made for this wonder chemical might be excessive and even dangerously shortsighted. Popular science magazines issued a call for "total war" against insects. The Department of Agriculture made DDT available to the public in massive quantities before manufacturers tested its potential danger to humans. Edwin Way Teale, a leading ornithologist and another of Carson's eminent supporters, wrote an article in the March 1945 issue of *Nature Magazine* warning that DDT might pose a threat to wildlife. Elmer Higgins, who first hired Carson and urged her to publish in the *Atlantic*, was one of the researchers at Patuxent. Together with biologist Clarence Cottam, yet another of Carson's scientific allies, Higgins sent a series of reports critical of DDT to Fish and Wildlife for her to edit. As with radar, Carson proposed to translate their reservations into terms a popular audience could understand. The *Digest*, however, had no appetite for scientific muckraking, especially in the summer of 1945. The triumph of science was the big story in a year in which the United States successfully detonated atomic bombs at Hiroshima and Nagasaki. When the *Digest* turned her down, she dropped the subject, even though her skepticism lingered.[11]

Despite her frustrations, Carson embraced the new opportunities her job offered. She traveled to research stations and nature preserves at government expense. At those sites, she occasionally found time for fieldwork, while making contacts with scientists doing oceanographic research. For the moment, her family life stabilized into manageable routines. Both of her nieces, Marjorie and Virginia, worked part-time jobs, though they continued to live at home. Maria Carson still ran the household, allowing Rachel the freedom to travel and socialize with friends. Even as she chaffed at the limitations of her job, Carson began to explore the background for what would become her next book. A

combination of personal and professional circumstances prepared her to fulfill her destiny as a writer.

Shortly after World War II, Carson proposed that Fish and
Wildlife publish a series of booklets, called Conservation in
Action, to introduce the general public to the national wildlife
refuge system. She wanted to promote public awareness of this
vital government program but also to overcome the human-
centered view of nature that most Americans held. She wrote, in
one booklet, that the survival of humans required "the preservation of the basic resources of the earth, which man, as well as animals, must have in order to live." Echoing the jeremiads of her
Presbyterian forebears, she warned, "Wildlife, water, forests,
grasslands, all are parts of man's essential environment; the conservation and effective use of one is impossible except as the others are also conserved." The refuges, as she explained, provided
for "the preservation of wildlife and wildlife habitat." Over the
years, conservationists, preservationists, and naturalists identified these unique habitats that they believed needed protection
from development.[12]

To gather material for this series, Carson traveled to sites
along the East Coast and west to Utah, Montana, and Oregon.
Conditions at the refuges were often primitive. Only modified
jeeps, swamp buggies, and open boats could traverse the roadless
terrains. At Chincoteague in Virginia, the refuge manager's boat
sometimes dropped Carson and her friend Shirley Briggs well
short of the beaches. The normally demure Carson could not

Despite frustrations with her job, Carson (sitting) had many opportunities to visit wildlife refuges, such as this one at Cobb Island, Virginia. (SOURCE: *Courtesy of the Rachel Carson Council.*)

avoid wading in the muck and mire to reach the shore. She welcomed the experience since Chincoteague was a bird-lover's paradise. Several flyways for Canadian geese, ducks, and other migratory waterfowl converged at that point, where the dunes and marshes afforded them a safe haven. Briggs noted with amusement the reaction of their fellow hotel guests when they "came lumbering through [the lobby], wearing old tennis shoes, usually wet, sloppy and be-smudged pants, various layers of jackets, souwesters, and toting all manner of cameras, my magnificent tripod, and Ray's binoculars."[13]

Whenever pressure slackened at Fish and Wildlife, Carson spent her free time reading about the oceans and collecting material for what she envisioned as a natural history of the seas. Work-related trips to coastal refuges, fish hatcheries, and fishery research centers afforded her additional opportunities to collect materials and make contacts. One trip to Parker River Refuge along the northern Massachusetts coast in September 1946 proved unusually fruitful. Her fieldwork was "full of command cars, marshes, mud, sand dunes, and Audubonites (amateur bird watchers)." At the end of each day, she felt "sunburned, black-and-blue, mosquito-bitten and weary." Still, she found time to search the records of the Massachusetts Audubon Society for data on waterfowl conservation at Parker River. More important, she managed a visit with Henry Bigelow, who having once headed the Woods Hole marine laboratories, now served as curator for the Museum of Comparative Zoology at Harvard. Carson wanted to learn all she could about Bigelow's current work and his views on oceanographic studies published during World War II. Her curiosity and command of the field must have impressed him because he became another of her many scientific supporters. With his encouragement, she left Massachusetts thinking more about how her book could broaden popular understanding of the oceans.[14]

By 1948 Carson's friends and colleagues sensed that the book project had matured into a serious endeavor. Shirley Briggs noted in March in her journal, "Ray is concocting a fine scheme whereby she writes a Bestseller—*Forever Amber* theme—perhaps using an Audubon Society background, and is so wealthy and can retire and write natural history." What Briggs wrote in jest proved closer to the truth than she could have imagined. Carson already envisioned an outline for the book she intended to write. In September 1948, she explained in a letter to William Beebe something about what she had in mind. "I am much impressed by man's dependence upon the ocean, directly, and in thousands of ways unsuspected by most people," she commented. "These relationships, and my belief that we will become more dependent on the ocean as we destroy the land, are really the theme of the book and have suggested the title, 'Return to the Sea.'" Her observation about destroying the land and about human dependence on the ocean probably seemed self-evident to a scientist such as Beebe. Yet its underlying ecological view of the interdependence of living creatures was still far outside the popular consciousness.[15]

Charles Alldredge, a friend and colleague who appreciated Carson's literary gifts, sensed that she was on the verge of something important. Alldredge also understood the many conflicting demands on her time and energy. He suggested that she could better concentrate on the new book if an agent handled the legal and business side of her work. After years of worrying about money, Carson was loath to share royalties with anyone, but her difficulties with various publishers persuaded her to accept Alldredge's advice and his list of possible names. On that list, she discovered Marie Rodell.[16]

Marie Rodell was an inspired choice, sophisticated and vivacious in ways Carson could never hope to be. The daughter of Russian Jewish immigrant parents, Rodell grew up in New York City, where her father made a comfortable living in the insurance business, while her mother taught school and wrote plays. As a young girl, Rodell attended the progressive School of Ethical

Culture in New York, learned to speak four languages fluently, and traveled widely. After graduating from Vassar in 1932, she worked as an assistant editor for a New York publisher and wrote novels and mystery stories in her spare time. A brief marriage to a failed writer ended with an abortion and a divorce.

Like Carson, Rodell became a victim of the publishing world's fraternity culture. When she lost her job to a man, she decided to strike off on her own as a literary agent. Her wide contacts, experience with the business side of publishing, and editorial skills, not to mention her own successes as a writer, made her an ideal agent. The two could not help but appreciate their shared ambitions as women striving to succeed in a male domain. Besides Rodell's wit and charm, Carson was also attracted to her toughness and personal integrity. Rodell always did what was best for her clients. For Carson, she offered a combination of shrewd business advice, critical judgment about writing, and warm friendship. More important, she recognized in Carson a writer of great promise.

Though five years younger, Rodell became another of the many female mentors who guided Carson's path to success. This friendship emerged at a crucial time because Rodell filled a void left by the unexpected death of Mary Scott Skinker. Carson remained in touch with Skinker after she left Washington to return to teaching. In November 1948, Carson learned that her friend was dying of cancer and borrowed money to be at her bedside in Chicago. To Rodell, Carson confessed she was "pretty well shot to pieces emotionally by my friend's illness and the tragic circumstances connected with it. . . ." Skinker died one month later. The loss of such a dear friend devastated Carson. Skinker had advised her at each step of her career—befriending her when she was an awkward student at Pennsylvania College for Women, drawing her into biology, nurturing her talents and ambitions, attracting her to Woods Hole, directing her into government, and shaping her growing ecological perspective.[17]

Rodell provided welcome emotional support, but as an agent her principal role was to help Carson realize the potential of *Return to the Sea.* Besides finding a publisher, Carson needed time off from Fish and Wildlife to concentrate on the book. That, in turn, required some outside source of funding to support her family. The sale of chapters to magazines was another possible source of income but only in the short run and only in a trickle. The promise of a flood of royalties lay several years away.

Rodell first freed Carson from any legal obligations to Simon and Schuster, then looked for a new publisher capable of promoting the book. Publishers generally make their decisions based on a writing sample and a detailed outline. Producing an outline posed no problem because Carson had already developed a plan. The sample chapter, however, came less easily. Over the summer of 1948 Carson had been researching a chapter that became "The Birth of an Island." As always, she read widely and consulted experts in the field, in this case oceanic geologists and tropical botanists. Her problem, as she confessed to Rodell, was "finding so much terribly interesting stuff that I can't stop researching." Having promised the island chapter by early September, Carson finally finished it at the end of the month.[18]

As pressure built, a productive dynamic emerged between author and agent. Rodell acted as editor, cheerleader, as well as legal and business adviser. Carson showered her with ideas for articles and with manuscript from work in progress. Rodell, in turn, understood both Carson's restless curiosity and her need for money, so she followed up only on the good ideas for articles, while keeping her client and friend focused on the task at hand. "I applaud your decision about no more articles until a hunk of the book gets done," she wrote the following November. "We do get distracted don't we?" Carson accepted the gentle chiding in the constructive spirit Rodell intended. After a series of rejections, Carson admitted that "these magazine things are a mistake, at least until the book has a solid start. So the next time I suggest one, please drop my letter in the nearest wastebasket."[19]

As an editor, Rodell spent little time on Carson's prose. Instead, she concentrated on making each piece more inviting to the popular audience Carson hoped to reach. That generally meant condensing, adding human interest, and finding details to enrich the narrative. More important, she gave Carson encouragement and emotional support. Carson always thought of herself as a slow writer. Certainly, she was a perfectionist, subjecting every draft to constant revision. Often, she became discouraged with her lack of progress and inability to meet deadlines. Her job compounded the problem. She frequently arrived at the weekend so exhausted that she could not face her own writing. A call to Marie (the two were quickly on a first-name basis) often buoyed her spirits. Rodell had an irrepressible and irreverent sense of humor. She once told Carson she had "sacrificed a black cock" and "put a hill of beans under the bed" before sending a piece to *The Saturday Evening Post*. She also assured Carson, "I think it reads very well."[20]

Carson needed such reassurance in the early months of their relationship. The first publisher to whom Rodell sent *Return to the Sea* rejected it as "too great an undertaking on the basis of the material submitted." Carson took the setback in stride, but it was followed by a series of rejections from magazine publishers. This was also the time at which Mary Scott Skinker died. The combination of pressure, disappointment, and emotional distress taxed her none-too-robust constitution. Even when Rodell left for a two-week vacation with her parents in Florida, she continued to offer Carson support and encouragement.

Over the next trying months, Carson made slow but crucial progress. A visit from Rodell to Maryland inspired a breakthrough. Carson discovered that one reason the book was going so slowly was that she had not written the first chapter. Without it, none of the other chapters fit into a larger whole. Once Carson conceived that chapter, she could "see all the others falling into line in a way they just refused to do up to now." To Rodell she remarked, "Absurdly obvious, isn't it?" After that epiphany, Carson produced four chapters, a revised island chapter, and a new outline. In the process, she discovered a personal voice. "At least what I am writing now is me," she confessed to Rodell, "even if somewhat different from what I thought it would be."[21]

As Carson's project matured, chance brought Rodell an overture from Oxford University Press. Philip Vaudrin, Oxford's editor, was eager to move the press out of its academic market and into trade publishing. An Oxford sales representative who heard about Carson from a bookstore owner in Washington, D.C., passed the information to Vaudrin. He contacted Rodell, who sent him the finished chapters and outline. Readers' reports confirmed Vaudrin's sense that he had a new author with a promising book. When Carson met Vaudrin in Washington, she "liked him immensely." They agreed on terms that included a welcome $1000 advance. Eager to move ahead, she refused Vaudrin's request to see each chapter in draft. "That, my dear" she confided to Rodell, "is a privilege only you can have." Vaudrin set March 1, 1950, as the delivery date for the manuscript.[22]

Facing a tight schedule, Carson needed free time more desperately than ever. Again, Rodell came to her rescue. She learned that the Saxton Foundation made awards to unrecognized writers of promise. Carson qualified on both accounts—she was a gifted writer struggling for recognition. And the money was critical if she had any hope of meeting her March 1 deadline. In her application to Saxton, she explained that she required at least four uninterrupted months to devote to the book and sufficient funds to provide for her niece in college, an ailing mother, and her own personal expenses. As justification, Carson laid out her vision for the book. Her passion for her subject came from her conviction "of the dominant role played by the ocean in the course of earth history" as well as her belief that "all life everywhere carries with it the impress of its marine origin." The audience for this book was anyone who like herself "has ever seen [the ocean] or has felt its fascination even before standing on its shores." That did not mean she would slight the scientific point of view. Exposing popular audiences to the most authoritative science was the signature of her work. Thus, she intended her book to be one "with which scientists will have no quarrel, and in which they may even find a fresh approach and a fresh interpretation of matters in a broad way familiar."[23]

Carson properly imagined herself in the role of those rare writers, such as the early ecologist Aldo Leopold, whose capacity for broad synthesis and clear exposition shaped their world's understanding of nature in ways most scientists seldom could. In an era of narrowly focused research, scientists had less capacity for grand theory. As Carson explained to Saxton, she hoped to provide "an imaginative searching out of what is significant in the life history of the earth's ocean," while at the same time addressing "questions thus raised in the light of the best scientific knowledge." What made her quest possible, she emphasized, was the enormous increase in the techniques for oceanographic research and advances in scientific understanding of the marine world.[24]

Having found a potential benefactor, Carson faced her research with new resolve. On an April trip to see Rodell in New York, she managed to spend some time with William Beebe, who persuaded her that an underwater dive would expand her knowledge of the oceans. As a result, Beebe made arrangements for her to dive in Bermuda "so that," Carson explained to Rodell, "I'll be sure of meeting the proper sharks, octopuses, etc." As it turned out, she had her "great undersea adventure" in Miami rather than Bermuda. That July, accompanied by Shirley Briggs, she toured Florida Bay, followed by a trip deep into the Everglades. Here, in this primal swamp, land and sea merged in the real world as well as in Carson's imagination. "There is the feeling that the land has formed only the thinnest veneer over the underlying platform of the ancient sea," she wrote in her field notes. "The feeling of space is almost the same as the sea, from the flatness of the landscape and the dominance of the sky."[25]

The ocean proved less welcoming. For three days Carson and Briggs headed out to the Keys in a small dive boat. For three days the wind blew and gathering storms made diving dangerous. On the fourth day, conditions remained forbidding, but the dive master allowed her a brief entry below the surface. Though she never left the boat's ladder, Carson did have the experience she most sought: a view of the ocean's surface from underneath it.

Whatever disappointment she felt at not exploring the underwater world more thoroughly was more than offset by a phone call from her mother, who informed her that she had won the Saxton fellowship. Writing to Beebe to thank him for his reference letter to Saxton and his sponsorship of her underwater adventure, she explained that while conditions for her dive "were far from ideal—water murky, the current so strong I could not walk around but hung to the ladder," she nonetheless recognized that "the difference between having dived—even under those conditions—and never having dived is so tremendous that it formed one of those milestones of life, after which everything seems a little different."[26]

Fellowship in hand, she pressed on to finish her book with a new urgency. Carson rushed home, repacked, and headed off to meet Rodell at Woods Hole for a trip on *Albatross III*, Fish and Wildlife's ocean research ship. The ship's destination was the Georges Bank, a rich fishing ground 200 miles off the North Atlantic coast, where the cold north winds collide with the warm breezes from the Gulf Stream. The trip allowed Carson to observe the shifting winds, currents, and fogs that fishermen had experienced for hundreds of years. Normally, government officials would not allow a single woman on board a ship with an all-male crew. Rodell's willingness to act as a "chaperone" solved the problem since, for some bureaucratic reason, they accepted two women but not one.

Albatross III was far from the ideal observation post. Long and narrow, with a high superstructure, the ship rolled like a canoe in the open ocean. The captain warned his guests that they would most likely be seasick, but they were not. Instead, the boat's fishing gear kept the two exhausted women up much of the night as it thumped and roared with the effect of two ships colliding. Physical discomfort could not chill Carson's fascination with what she was seeing. The nets gathered all manner of creatures from depths of as much as 600 fathoms (3600 feet or about two-thirds of a mile): "Scores of different species of fishes were brought up to give their own mute testimony as to who lived down in the undersea world of Georges Bank."

Carson saw, too, a darker side to this work. From those depths "there is a strange effect, caused by the sudden change of pressure," she observed. "Some of the fish become enormously distended and float hopelessly on their backs." Carrion generally attracted sharks. Carson was disturbed to note that while she saw a strange beauty in these predators, some of the crew shot them for sport. Still, she marveled at the seascape *Albatross III* encountered. "When I stood on the after deck on those dark nights, on a tiny man-made island of wood and steel, dimly seeing the great shapes of waves that rolled about us," she recalled, "I think I was conscious as never before that ours is a water world dominated by the immensity of the sea." The ability to transfer that sense of awe from what she saw and imagined to the printed page distinguished Carson among naturalist writers.[27]

During their many conversations aboard *Albatross III*, Carson talked to Rodell about other books she might write. Rodell later presented one of those ideas to Paul Brooks, an editor for

Houghton Mifflin. In preparing to meet Carson, Brooks read a prepublication copy of *Return to the Sea*. "I liked it so much," he told Rodell, "that I read every word with pleasure and admiration." Brooks then wondered if Carson had other projects in mind and added that he had an idea for a book that might interest her. When they met several weeks later, an instant friendship was born. Brooks was a New England patrician with a warm, human side. The two shared interests in both ornithology and natural history. They had numerous friends and acquaintances in common, chief among them Roger Tory Peterson, the renowned bird illustrator with whom Carson served on the board of Washington's Audubon Society. Brooks edited Peterson's widely popular field guides to the birds of North America.[28]

Using Peterson as a model, Brooks proposed that Carson write a book to educate the public about the ecology of common shore creatures. The idea had come to him in a comedy of ecological errors. Among his editors at Houghton Mifflin was Rosalind Wilson, the daughter of the famed literary critic Edmund Wilson. When Wilson joined a group of her father's bohemian guests at his cottage in Provincetown on Cape Cod, a major storm interrupted their weekend revelry. The morning after, they walked along the beach assessing the damage. There, the bohemian band discovered what appeared to them as masses of horseshoe crabs strewn helplessly on their backs. Moved by humane concern for the plight of their fellow creatures, they dutifully tossed the crabs back into the water, only to discover later that the crabs were mating. What they had done in sympathy had likely interrupted the crabs' reproductive cycle. If these sophisticates could be so ignorant of basic biology, Wilson wondered, what could one expect of the broader public? Brooks agreed that a seashore guide might serve a useful purpose. He had no difficulty persuading Carson because she already recognized the appeal of such a book. The two agreed that she would write it after she finished *Return to the Sea.*[29]

Despite her progress, occasional leaves from her job, and the promise of a new book contract, Carson struggled through the last months of 1949 and the first half of 1950. The family moved in September, niece Marjorie suffered a new round of illnesses (complicated by her diabetes), and office work continued to eat into Carson's writing time. In addition, Carson grew increasingly dissatisfied with *Return to the Sea* as a title. She believed, and Rodell concurred, that it did not adequately capture the sweep of what she was writing. Above all, she wanted to avoid a title that seemed too "textbookish" or academic. *The Story of the Ocean* struck her as appropriate but boring. *Story of the Sea, Empire of the Sea,* and *Sea Without End* all failed to capture the essence of the book. Her friends, noticing her frustration, sought to cheer her with such irreverent alternatives as *Carson at Sea* and *Out of My Depth.* Rodell's failure to find magazines to serialize the book or publish chapters only added to her on-going financial anxieties. The Saxton Foundation compounded those worries by reducing her fellowship by the $1000 she received as her advance from Oxford.[30]

As the March 1 deadline loomed, Carson was determined to finish but confessed to Rodell "I feel that I'd die if I went at this much longer." Sensing her client's distress, Rodell negotiated a new deadline for the book. As soon as the pressure eased, Carson's life brightened. She sent her chapters out for review by scientists on whose work she drew. All were enthusiastic and generous in their praise. Harvard meteorologist Charles Brook congratulated her, "on a very interesting chapter." She in turn thanked him for his careful review. He had "corrected the errors of phraseology," which she confessed resulted "from dealing with a subject as foreign from my training as meteorology." No response meant more to her than that from Henry Bigelow who "read every word . . . with great pleasure" and predicted that "your book will stand for a long time."[31]

Better yet, several journals, including the respected *Yale Review,* offered to publish some of her work. Best of all, *The New Yorker* expressed interest in condensing several chapters. Not only did *The New Yorker* set the tone for the sophisticated reading public; it paid well. Exposure in its pages virtually guaranteed brisk book sales. To top things off, Carson returned to a title she had considered once before: *The Sea Around Us.* It hit precisely the right note. By late June 1950 Carson delivered the manuscript, although she continued to revise it through galleys and page proofs.

It took a full year for *The Sea Around Us* to reach the bookstores. In that time, interest in the book began to build. William Shawn, the editor of *The New Yorker,* asked for nine chapters and added that he would personally condense them. They would be published as a three-part series called "Profile of the Sea," for which Carson would receive $7200, far more than she had ever earned from writing and more than her annual salary at Fish and Wildlife. Soon after, *The Yale Review* offered $75 for "Birth of an Island." The tropical botanist who refereed her article for *The Yale Review* called it "the finest account of the creation and colonization of an oceanic island" he had ever seen. While the *Review* paid only a nominal sum, publication in September 1950 made Carson eligible for an annual prize awarded by the American Association for the Advancement of Science (AAAS). In addition, *Science Digest* paid her a handsome sum for another chapter, "Wealth from Salt Seas."[32]

Despite these encouraging signs, Carson's spirits sagged, as they often did after periods of intense creativity. She called it "nervous exhaustion" yet had trouble sleeping. "I seem quite incapable of unwinding and really relaxing," she wrote her old friend Maria Leiper from Simon and Schuster, "but perhaps it is as well—books don't get written that way." In September a physical checkup discovered a small tumor in her left breast. (A cyst in the same breast had been removed four years earlier.) Carson casually informed Rodell "the doctor thinks I should get rid of [it], and I suppose it's a good idea." After that, she reported, she was "ordered to take a vacation as soon as the hospital episode is out of the way." Not until early October did she get away to Nags Head on the North Carolina shore.[33]

Walking along the expanses of Nags Head beaches, she reflected on the transitory nature of life and her own plans as a writer. "Time itself is like the sea, containing all that came before us," she observed as waves flowed over the shore, "sooner or later sweeping us away on its flood and washing over and obliterating the traces of our presence, as the sea this morning erased the foot-prints of the bird." She also thought about the shore book she intended to write. At first, she conceived of it as a practical guide to identify flora and fauna along the seashore. Now, she began to imagine it as a companion to *The Sea Around Us*. In that book, she focused on the physical character of the oceans; in this one, she wanted to tell "the story of how that marvelous, tough, vital, and adaptable something we know as life has come to occupy one part of the sea world and how it has adjusted itself and survived despite the immense blind forces acting on it from every side." If oceanography defined *The Sea Around Us*, ecology would inform what she was calling *A Guide to Seashore Life on the Atlantic Coast*.[34]

But first things first. Before Carson could begin her next book, she had to attend to the final details of *The Sea Around Us*. Anxious moments awaited her. Ten years earlier, *Under the Sea-Wind* stirred enthusiastic responses, only to fade into obscurity under the winds of war. Carson was determined that would not happen this time. She worried the book might be "dismissed as another introduction to oceanography," as had happened to another recently published book. Her anxieties rose when she saw the first mock-up. She had warned Vaudrin "that every care must be taken to avoid the physical appearance of a textbook." Yet, to her professionally trained eye, that was just what Oxford was doing. She objected to the typeface the designers chose for chapter and section headings and suggested alternatives. Vaudrin quickly reassured her that he understood her concerns. But in February he left Oxford, forcing Carson to deal with the president, Henry Walck, whose background was in accounting, and with the people in marketing, who were far more familiar with academic than trade publishing.[35]

Then there was Korea. Might war once again distract the public from her book? North Korean troops swept into the south in June of 1950. General Douglas MacArthur organized a brilliant counteroffensive to rout the invaders. By September, American forces crossed the 38th parallel as MacArthur confidently promised to have the boys home by Christmas. That was before Chinese forces crossed the frozen Yalu River on the border of China and North Korea on November 26. They inflicted one of the worst defeats in American military history. Carson wondered if this signaled the beginning of a larger war between China and the United States. The combination of the war and massive new defense spending led to rising prices and shortages of key materials—among them paper. At Fish and Wildlife, key personnel anticipated return to military service. Each day brought reassignments and a growing possibility the agency would be moved out of Washington. At *The New Yorker,* William Shawn decided to hold publication of Carson's chapters until he had a clearer sense of how the war was going. Carson had every reason to fear that this would be 1941 all over again.

No matter how justified, Carson's war worries proved groundless. By January, American forces had driven the Chinese back to the 38th parallel. More important, President Harry Truman and British Prime Minister Clement Attlee agreed they would not allow Korea to become a general war. By spring, the fighting had settled into a bloody stalemate.[36]

As her anxieties over Korea ebbed, Carson faced a new problem, one she had never anticipated—success and the fame that came with it. December 1950 brought an omen. The AAAS Westinghouse Prize Committee chose "Birth of an Island" for its science writing award and a prize of $1000. Despite a crowded schedule, she had to travel to Cleveland to accept it, though to her relief she was not expected to make an acceptance speech.

Carson had little experience with or taste for public speaking. She was a private person, happiest with her writing and the company of friends and family. But her privacy was, as it turned out, the first casualty of fame.

In February, Rodell sent more good news. The Book-of the-Month Club was considering *The Sea Around Us* as a monthly selection. Unable to contain her joy, Carson dragged a colleague into a phone booth to tell him the news. Selection, she told him confidentially, would earn her enough money to leave the agency and live as a writer. March brought glad tidings of another sort. On a whim, Carson had applied for a Guggenheim Fellowship and won. The award allowed her to take a year's leave from Fish and Wildlife. Rodell called several days later to say that Book-of-the-Month indeed had made *The Sea Around Us* an alternative selection—not quite the brass ring but promising enough to satisfy Carson. Still, she continued to deluge Rodell and Oxford with suggestions on how best to market and promote the book.[37]

Carson need not have worried. *The New Yorker* did far more for the success of her book than she imagined. William Shawn spent ten days with her condensing the chapters into a three-part series that he began publishing on June 2. Working with Shawn, a superb editor, gave Carson a sense of literary craftsmanship. And working with Carson persuaded Shawn he was publishing a writer of the first rank. No sooner had the series opened to widespread acclaim than Rodell learned that the *Saturday Review of Literature* planned to feature Carson on the cover of its July 7 issue. Like *The New Yorker*, *Saturday Review* was a magazine that the literati and the reading public took seriously. An elated

Carson asked Rodell, "Are they doing me against a background of squids, spouting whales, etc.?"

One early June morning had brought recognition of yet another kind. Alice Roosevelt Longworth called to say she had been given an advance copy of *The Sea Around Us* and had stayed up all night reading it, not once but twice! She told Maria Carson, who answered the phone in Rachel's absence, it was "the most marvelous thing she had ever read." Longworth was not simply the daughter of Theodore Roosevelt; she was also the doyenne of Washington society and a razor-sharp critic. Her enthusiasm was a useful predictor of how a wider public might receive the book.[38]

On July 1, *The Sea Around Us* was featured in the prestigious *New York Times Book Review*. From that point on, reviewers, whether writing for scientific or literary journals, praised Carson for the high quality of her science and her prose. The *Scientific Monthly* noted, "It is singularly refreshing to find a popular book on a scientific subject which can carry the reader so gently and pleasantly through the mazes of science and yet maintain such a high degree of accuracy." Similarly, the *Atlantic Monthly* praised her as a marine biologist who had written "a first-rate scientific tract with the charm of an elegant novelist and the lyric persuasiveness of a poet." A scientist who had himself written a book on the oceans appreciated that, rather than focusing on the more popular biological aspects of the oceans such as sharks, whales, and giant kelp forests, *The Sea Around Us* took on the far more difficult task of describing the "physical, geographical, and esthetic aspects of its subject." A panel of seventy leading authors picked *The Sea Around Us* as the best book of the year, ahead of

Catcher in the Rye. Honorary degrees, awards, and prizes were showered on Carson. To top off her success, she won the most prestigious prize of all, the National Book Award for nonfiction.[39]

At age 44, Rachel Carson realized her lifelong dream as commercial success followed literary recognition. Oxford went through fifteen printings in the first two months. Sales exceeded 100,000 by early November and climbed still higher as Christmas approached. *The Sea Around Us* spent a record eighty-six weeks

Among the many awards she received for The Sea Around Us, *none carried more prestige than the National Book Award. Here, at the 1951 ceremony, she is seated to the right of the novelist James Jones and opposite the poet Marianne Moore. (SOURCE: Courtesy of the Rachel Carson Council.)*

at the top of the *New York Times* bestseller list (beating out Thor Heyerdahl's *Kon-Tiki*). It was translated into thirty-one foreign languages, and its first edition finally sold over 1.3 million copies. Once she was financially secure, and with encouragement from Marie Rodell, Carson resigned her position at Fish and Wildlife in May 1952. Now she could write full-time.

Carson's success was no accident. Through talent and perseverance she made her dream a reality. *The Sea Around Us* had the power to fascinate, to inform, and to inspire. Turn to the section on "tsunami" and you find a cogent explanation of the phenomenon that in 2004 devastated much of South Asia. Where did the moon come from, she asked her readers, then offered a lucid explanation based on the geology and physics of her time. When the earth's surface was forming, powerful solar tides developed so much momentum that they tore away a great mass of the earth's surface and threw it into space. That mass became the moon and the "scar or depression left behind" now holds the Pacific Ocean. Not only did Carson inform, she invited her reader to view worlds unseen. "Imagine a land of stone, a silent land, except for the sound of rains and winds swept across it," she wrote of the earth in its formative phase, "for there was no living voice and no living thing moved over the surface of the rocks."[40]

For readers whose imaginations needed prompting, she provided clarifying analogies. How, for example, to understand the declining plant life as one descends deeper into the oceans? Carson painted a word picture of the "sunless sea" and "hidden lands" where one discovers "undersea mountains, with their sheer cliffs and rocky valleys and towering peaks." These were seascapes unfamiliar to most readers. So Carson suggested that they think of the regions above the timberline "with their snow-filled valleys and their naked rocks swept by the winds." The oceans, she then pointed out, have an "inverted timber line." "The slopes of the undersea mountains," she explained, "are far beyond the reach of the sun's rays, and there are only bare rocks, and in the valleys, the deep drifts of sediments that have been silently building up for millions upon millions of years." Lest readers have the illusion that this was a lifeless place, Carson assured them otherwise. "Biologically, the world of the continental slope, like that of the abyss, is a world of animals," but because no plants live there, it is also "a world of carnivores where each creature preys upon the others."[41]

The sea world Carson unveiled was a place worthy of science fiction—home to all manner of exotic creatures and unexpected sights and sounds. Too many people live with the illusion, she noted with disdain, that "any light not of moon or star or sun" must have a human origin. Not so, for "the deep sea has its stars, and here and there an eerie and transient equivalent of moonlight, for the mysterious phenomenon of luminescence is displayed by perhaps half of all the fishes that live in dimly lit or darkened waters." Here were fish with lanterns they "could turn on and off at will," fish with rows of light to signal both friend and enemy, and the deep-sea squid that, unlike its dark ink-squirting, shallow-water cousins, ejected a luminous cloud. The notion that these depths were a silent place, Carson assured her readers, was "wholly false." Hydrophones extended on cables to great depths recorded "strange mewing sounds, shrieks, and ghostly moans," the source of which remained a mystery. And on rare occasions, the depths gave up some of their strangest denizens. In 1938, fishermen off the southeast tip of Africa discovered in their nets a strange bright blue fish, genus *Latimeria,* that scientists assumed to have been dead for 60 million years! Then came the frill shark, a creature seemingly more like a reptile than a fish, whose many gills and three-pronged, briar-like teeth distinguished it from modern sharks and linked it to creatures from 25 to 30 million years earlier.

For most readers in 1951, the oceans remained as mysterious as these creatures captured from its sunless depths. Much of what people knew or thought they knew came from the myths of the ancient worlds, the tales of seafarers, or stories such as Jules Verne's wonderful *Twenty Thousand Leagues Under the Sea.* Carson offered her reader instead a narrative based not on myth but on science and yet with all the wonder and mystery still intact. Was it true that ships drifting into the Sargasso Sea were destroyed by sea monsters or trapped forever by giant weeds as Columbus's crew feared? Was there a lost continent of Atlantis? Did terrifying monsters live in the ocean depths as Verne portrayed? The answers according to Carson were no, no, and possibly. After explaining the source and biology of the weeds and creatures of the Sargasso, Carson concluded, "the gloomy hulks of vessels doomed to endless drifting in the clinging weed are only the ghosts of things that never were."[42]

As for "lost continents," especially Atlantis, which Plato had located beyond the "Pillars of Hercules," Carson noted that such legends "persistently recur like some deeply rooted racial memory in the folklore of many parts of the world." Science, she argued, proved it impossible that humans ever witnessed Atlantis before it was "lost." The most likely candidate, the Atlantic Ridge, splits the Atlantic Ocean from north to south. Core samples taken along its length indicated that the ridge had been submerged for some 60 million years, while hominids have walked the earth for only 1 million.

Carson granted that something did exist that might have given rise to the myth. The Dogger Bank, a plateau the size of Denmark lying some 60 feet beneath the surface of the North Sea, was the most likely subject. Just 30 to 40 thousand years ago, the area was dry land alive with forests, bears, wolves, and even the wooly rhinoceros and mammoths. Primitive humans hunted there as well. As the world warmed and the glaciers melted, the sea turned the plateau into an island and then immersed it altogether. Carson speculated that some people escaped and possibly "communicated this story to other men who passed it down through the ages until it became fixed in the memory of the race." Modern people know something of the story because fishing boats have dragged up in their nets bones, trees, moorlog, and the crude stone weapons early hunters left behind.[43]

Among the great behemoths of the oceans, Carson informed her readers, were the sperm whale and the giant squid—terrors in the imagination of landlubbers and deadly enemies of each other. Though the whale is a mammal, it can dive to depths as great as 3000 to 4000 feet. There, it sometimes locks in mortal combat with giant squid that, with their tentacles, may stretch to lengths of 50 feet. Whalers reported that many a sperm whale showed the interlaced scars left by the squid's suction cups.

Certainly, creatures such as these could inspire the myths of monsters and sea serpents. Carson set their story within the real-world ecology of their species. How do whales, for example, survive the extreme pressures of the deep and avoid caisson disease (the bends) when rapidly ascending to the surface? Carson noted that "the real mystery of sea life in relation to great pressure is not the animal that lives its whole life on the bottom bearing a pressure of perhaps five or six tons, but those that regularly move up and down hundreds or thousands of feet of vertical change." She speculated that in discovering the rich food sources in the ocean depths, whales developed the physical capacity to survive the extreme conditions.[44]

Carson broke her story into parts: creation; surface waters and sunless depths; geography of the sea floor; winds, tides, and currents; resources; and effects on climate. Her larger purpose was not to dissect but to give a sense of the whole. All life, including human life, bears the mark of its marine origins; all life depends on the seas; and in the end, as lands erode, all matter, organic and inorganic, returns to the sea—"to Oceanus, the ocean river, the ever flowing stream of time, the beginning and the end." To tell this story, Carson drew heavily on science. Indeed, in many ways, *The Sea Around Us* was a celebration of science and its practitioners. Carson not only reported what scientists had discovered about the sea but also explained how they came to know those things and why some explanations remain contested and some mysteries remain unsolved.[45]

In plumbing the depths of the oceans, Carson asked her readers to think ecologically. Hers was a biocentric view of nature, informed by a Darwinian approach to evolution, in which humans were just another species, albeit a dangerous and often destructive one. Inevitably, her Darwinian sensibilities provoked a challenge from those who saw the hand of God alone in creation, a response Carson took seriously, though she was by this time no longer a practicing Christian. When James Bennet, a New York attorney and amateur geographer, implied she was an atheist for failing to acknowledge God in the creation, Carson replied at length. "As far as I am concerned, however," she told Bennet, "there is absolutely no conflict between a belief in evolution and a belief in God as the creator." Evolution was the evident "method by which God created and is still creating life on earth." That process was, for her, "so marvelously conceived" that it only enhanced her "reverence and awe" for both the Creator and evolution.[46]

Carson's approach to ecology paralleled the work of Alfred North Whitehead, an English philosopher who settled in the United States in the 1920s and one of the scientists who came to be known as the "Ecology Group" at the University of Chicago. Whitehead's views of science and nature accorded neatly with those of someone raised, as Carson was, on the nature studies movement. Whitehead rejected what he saw as the central weakness of science over the past three centuries—its tendency to reductionism. Scientists sought to explain the physical world by dissecting it and separating it into its component elements. What was lost was a sense "of the organic unity of the whole." Everything in nature is linked to everything else, Whitehead argued, in what he, like Carson, saw as an intricate web of life. Where science tended to treat nature in mechanistic terms, as if it functioned like a machine, Whitehead stressed its organicism. The human body served him as a more appropriate analogy.

Further, to treat nature as a machine stripped it of any notion of value, ethics, or aesthetics. By contrast, organicism, as the proper study of nature, returned moral values to the scientific enterprise, Whitehead believed. A science that properly stressed the central idea of relatedness taught humans that interdependence was the essence of their being. The example he offered was the Brazilian rain forest. In it, a single tree was subjected to "all the adverse chance of shifting circumstances." Only under exceptional conditions, such as human cultivation, could that single tree flourish, which in ecological terms means to survive and reproduce. But in nature trees flourish as a collectivity, or what we call a forest. "Each tree may lose something of its individual perfection of growth," he conceded, "but they mutually assist each other in assuring the conditions for survival." Together they shade the soil, protect its vital microbes, preserve its moisture, and prevent it from washing away. As a consequence, "a forest is the triumph of the organization of mutually dependent species."

The Ecology Group gave this organicism ideological significance. To them, society, like nature, depended on the participation of individual parts within a complex whole, what Carson and others referred to as the "web of nature." That led them to conclude, "The individual metazoan, the infusorian population, the ant colony, the flock of fowl, the tribe, and the world-economy are all exemplifications of nature's grand strategy." That strategy was evolution toward greater interdependence and cooperation. The rise of totalitarianism in the late 1930s and 1940s drained much of the enthusiasm for this vision of an interlocking world. In time, the Ecology Group came to fear that cooperation could produce the kind of suffocating conformity characteristic of the totalitarian Nazi and Soviet states. By the 1950s, at the height of Senator Joseph McCarthy's anti-Red crusade, the ideal of individuals as subservient to interdependent communities smacked dangerously of Communism. Some members of the group began to look for an integrative model in which difference and diversity could survive. But before they could reformulate their ideas, the group dissolved and its ideas ceased to influence ecological thought.[47]

There was an exception. In *The Sea Around Us* and the books that followed, Carson embraced the cooperative ideal. She could do so because she had little concern with the political implications of ecology for human society. The moral condemnations that appeared occasionally in her writing had ecological, rather than political, import. She saw herself as a spokesperson for nature, a writer who promoted a sense of wonder and curiosity at the marvels of the living world. Scientists applauded her as someone who made their work accessible but not as a theorizer. As an advocate for nature, Carson did not want to reform humankind in the manner of the Ecology Group. Instead, she wanted to protect nature from degradation by humans.

Nowhere was that more evident than in the conclusion to her chapter "The Birth of an Island." In it, she lamented humankind's wanton destruction of isolated island habitats populated by unique species. In a more reasonable world, by which she meant a more biocentric one, humans should have preserved these oceanic habitats "as natural museums filled with beautiful and curious works of creation, valuable beyond price because nowhere in the world are they duplicated." Instead, as a fellow scientist observed,

"The beautiful has vanished and returns not." That message may have been lost on many of her readers in the 1950s, but it resonated with a new generation a decade later.[48]

NOTES

1. Mary Scott Skinker to Carson, December 9, 1941, RCP-BLYU.
2. Brooks, *House of Life*, pp. 70–71; Carson, speech to Theta Sigma Phi, April 21, 1954, RCP-BLYU.
3. Carson to Sonia Bleeker, March 3, 1942, RCP-BLYU.
4. Ibid.; Brooks, *House of Life*, p. 73.
5. Brooks, *House of Life*, p. 78.
6. Interview with Shirley Briggs cited in Lear, *Rachel Carson*, p. 125.
7. *Reader's Digest*, August 1945; see also Carson to Ada Govan, February 15, 1947, RCP-BLYU, and Brooks, *House of Life*, p. 78.
8. Carson to Quincy Howe, May 31, 1944, RCP-BLYU.
9. Carson to William Beebe, October 26, 1945, RCP-BLYU.
10. Carson to Harold Lynch, July 15, 1945, RCP-BLYU.
11. Dunlap, *DDT,* pp. 63–75.
12. Carson, *Guarding Our Wildlife Resources, Conservation in Action* (Washington, D.C., GPO, 1947). The link between Carson's view of nature and her Presbyterian roots is developed by Mark Stoll in two forthcoming essays, "Creating Ecology: Protestants and the Moral Community of Creation" and "Rachel Carson: Nature and the Presbyterian's Daughter," provided to me by the author.
13. Lear, *Rachel Carson*, p. 133.
14. Ibid., pp. 138–139; see also Carson to Briggs, September 25, 1946, RCP-BLYU.
15. Lear, *Rachel Carson*, pp. 145 and 148, and Carson to Beebe, September 6, 1948, RCP-BLYU.
16. Brooks, *House of Life*, p. 114; Lear, *Rachel Carson*, pp. 148–149.
17. Carson to Marie Rodell, December 15, 1948, RCP-BLYU.
18. Carson to Rodell, August 11, 1948, RCP-BLYU.
19. Rodell to Carson, November 15, 1948, and Carson to Rodell, January 9, 1949, RCP-BLYU.
20. Lear, *Rachel Carson*, pp. 156–159; Rodell to Carson, November 4, 1948, RCP-BLYU.

21. Carson to Rodell, January 24, 1949, RCP-BLYU; Brooks, *House of Life*, p. 114.

22. Carson to Rodell, June 8, 1949, RCP-BLYU.

23. Carson, application to the Saxton Foundation, May 1, 1949, RCP-BLYU.

24. Ibid.

25. Carson to Rodell, April 10, 1949, Carson to Beebe, April 1949, and Carson, field notes, "Tamiami Trail"—all in RCP-BLYU; see also Lear, *Rachel Carson*, pp. 166–167.

26. Carson to Beebe, August 26, 1949, RCP-BLYU.

27. Carson, "Aboard the Albatross," in Brooks, *House of Life*, pp. 115–119.

28. Lear, *Rachel Carson*, pp. 173–174.

29. Ibid., pp. 173–174; Brooks, *House of Life*, p. 153.

30. Carson to Rodell, undated, RCP-BLYU.

31. Carson to Charles Brooks, April 9, 1950, Brooks to Carson, March 22, 1950, and Carson to Rodell, March 16, 1950—all in RCP-BLYU.

32. Lear, *Rachel Carson*, pp. 183 and 177; Carson to Edwin Way Teale, September 19, 1950, RCP-BLYU; Brooks, *House of Life*, pp. 123–124.

33. Carson to Maria Leiper, October 26, 1950, and to Rodell September 10, 1950, RCP-BLYU.

34. Brooks, *House of Life*, pp. 152 and 154–155.

35. Carson to Philip Vaudrin, December 1950, and Vaudrin to Carson, December 21, 1950, RCP-BLYU.

36. For an able discussion of the Korean War, see Walter LaFeber, *America, Russia, and the Cold War, 1945–2002*, 9th ed. (New York, McGraw Hill 2002), pp. 118–129.

37. Rodell to Carson, February 14, 1951, RCP-BLYU.

38. Carson to Rodell, May 31, 1951, and June 14, 1951, and to Catherine Scott, June 14, 1951, RCP-BLYU; Lear, *Rachel Carson*, p. 198.

39. Reviews all collected in RCP-BLYU.

40. Carson, *The Sea Around Us*, pp. 21, 24, and 119–121.

41. Ibid., pp. 66 and 72.

42. Ibid., p. 40.

43. Ibid., p. 76.

44. Ibid., pp. 56–57.
45. Ibid., p. 196.
46. James Bennet to Carson, September 19, 1952, and Carson to Bennet, November 1, 1952, RCP-BLYU. On Carson's religious views, see also Lear, *Rachel Carson*, pp. 227–228, 444, and 544, n.78, and Stoll, "Rachel Carson," pp. 8–9, work in progress.
47. Worster, *Nature's Economy*, pp. 316–332.
48. Carson, *The Sea Around Us*, pp. 84–96.

· Three ·

FALL

The Fullness of Life:
From *The Edge of the Sea* to DDT

THE SUCCESS OF *THE SEA AROUND US* CREATED NEW OPPORTUNIties for Rachel Carson. Royalties freed her from the financial worries that had dogged her from childhood. With an independent income, she left government service and became a full-time writer, sought after by publishers with proposals for new projects. She celebrated her success with a brand new, two-tone Oldsmobile. Carson was freer, richer, and more productive than ever.

Such happiness was never complete, freedom never total. Having no idea that *The Sea Around Us* would sell so well, she had already contracted with Houghton Mifflin to do *A Guide to Seashore Life on the Atlantic Coast.* Deluged with speaking requests, showered with awards, and drowning in correspondence, she discovered that even though she had the means to concentrate on writing, she did not have the time.

Nor could Carson ignore her family responsibilities. She worried about her mother, who devoted herself to Rachel's success and managed her domestic life even as she grew older and more frail. By the time Rachel could provide financial security, Maria Carson was 83. With age came an increased need to remain at the center of Rachel's life. When Rachel traveled, Maria wrote her almost every day. On one trip, she posted her first letter just two hours after Rachel had departed. Some correspondence arrived even to the wilds of Montana, stamped "special delivery," to make sure her letters arrived promptly.

Dependence on Rachel led Maria to interfere with her daughter's friendships. No one had been more important to Rachel than Marie Rodell. Maria was jealous of the intimacy that developed between author and agent. She never welcomed Rodell during her visits. For her part, Rodell understood that Maria imposed on Rachel's writing time as their roles reversed and daughter began to care for mother. She urged Rachel to hire domestic help, but Maria resisted the idea of sharing her household duties with another woman. She insisted, too, on being present whenever Rachel entertained. That sometimes meant relationships suffered from lack of privacy since Rachel and her friends could seldom spend time alone together. Despite urging from Rodell and other well-meaning friends, Rachel could not easily bring herself to confront her mother over her increasingly unreasonable demands.

Trouble for Carson's niece Marjorie only made matters worse. Just as Rachel was beginning to enjoy her good fortune, she learned that Marjorie had been involved with a married man by whom she was pregnant. In the 1950s, out-of-wedlock pregnancy brought social disgrace. Given Marjorie's diabetic condition, an abortion was out of the question. Carson and her mother did all they could to shield Marjorie and protect her health. Somehow they managed to placate Rachel's older brother Robert, a sanctimonious soul, who saw his niece's condition as a mortal sin and viewed her as a family disgrace. Rallying around Marjorie, Rachel and Maria arranged for her medical care and kept her condition secret outside of the family until the baby was born.

The one person in whom Carson did confide was Rodell. Rachel felt that someone needed to understand why, at the point of her greatest success, she chose to travel for research and turned down several opportunities to promote her book. To others, such as the sales department at Oxford University Press, Carson explained that her doctor insisted she reduce the many obligations that were undermining her health. In the meantime, Maria Carson managed the domestic details of their deception. Roger Christie, her great-grandson and Carson's grandnephew, was born on February 18, 1952. Though Carson loved Roger and cared for him when his mother could not, she later recalled this episode as "a private tragedy" that had "blotted out" all the excitement of her newfound fame. "If ever I am bitter," she later admitted to her friend Dorothy Freeman, "it is about that."[1]

Bitterness was not in Carson's nature. After the publication of *The Sea Around Us*, she found her life full and rewarding in new ways. For one, she indulged her love affair with the Maine coast. She first discovered its splendor in the summer of 1946 when she and her mother rented a cottage on the Sheepscot River, west of Boothbay Harbor. The only sounds they could hear were "gulls, herons, and ospreys, sometimes the tolling of a bell buoy, and— when the wind [is] right, the very distant sound of surf," she told a friend. All that Carson loved in nature was at her doorstep—

birds, woods, and tide pools. She confessed at that time her "greatest ambition is to be able to buy a place here" where she could spend her summers.[2]

The Sea Around Us made that ambition a reality. In September of 1952, eight months after Roger was born, she wrote to Rodell to explain her late return from Southport, Maine: "I am about to become the owner (strange and inappropriate word) of a perfectly magnificent piece of Maine shoreline, and how by June I am to have a sweet little place of my own built and ready to occupy!" The site was not far from her first Maine cottage. It "overlooks the estuary of the Sheepscot River, which is very deep, so that sometimes—you'll never guess—whales come up past the place, blowing and rolling in all their majesty!" The lot she acquired measured 140 by 350 feet and was covered with evergreens and other trees. Locals called the nearby bluff "Dogfish Head," and it was near there she built her cottage. It remained her sanctuary for the rest of her life. From her porch she could climb down the bluff to the tide pools she loved to explore or wander into the woods behind it to observe the birds and to reflect.[3]

Only gradually did Carson recognize that the acquisition of this property coincided with a slow but pronounced change in her relationship with Rodell. Prior to that summer she wrote her friend regularly about personal matters as well as business. She regretted long absences and occasionally arranged to vacation with her. When she traveled to New York, she looked forward to nights in Rodell's sparkling company, even if her sofa was a bit lumpy. When Rodell's schedule brought her to Washington, Carson encouraged her to stay in Silver Spring. But from the winter of 1952 on she wrote Rodell less often and their exchanges

Carson's home on the Maine shore would be her refuge for the remainder of her life. Access to the sea and spectacular views made her feel close to the world she loved. (SOURCE: *Stanley Freeman, Jr. Permission granted by the Freeman Family Collection.*)

dealt largely with business, even though the tone remained warm and friendly.[4]

At this time, too, Carson found two new friends who filled roles Rodell had once played. The first was her new editor at Houghton Mifflin, Paul Brooks. Having signed Carson to do a seashore book, Brooks became someone to whom she now turned for advice and encouragement. She sensed immediately that he understood exactly what kind of book she had in mind. "Your general aims and mine are the same," she told him. When Philip Vaudrin at Oxford once asked to see draft chapters of *The Sea Around Us*, Carson declined and granted Rodell alone the privilege of seeing drafts. Carson harbored no similar compunction about Brooks. He became as much a part of her creative process as Rodell.

In June 1953, with the book well behind schedule, she sent Brooks a copy of a chapter on the Florida Keys. In writing it, she realized she had "been trying for a very long time to write the wrong kind of book." Rather than cataloguing the varieties of shore life, she now planned to group them ecologically. Her new insight gave the book thematic unity but also meant major rewriting. A sympathetic Brooks then revealed his deft editorial hand: "It requires quite an act of artistic creation to incorporate such a mass of facts into a pattern—or perhaps it would be more scientific to say to reveal the pattern under the mass of facts," he told his harried author. "In any case, that seems to be what you are doing here, and it is good."[5]

Shortly after reorganizing her shore guide, Carson headed off to Maine. There, she spent her first summer in the new cottage she had named "Silverledges." Among those who sent a note of welcome was her neighbor, Mrs. Stanley Freeman. Dorothy Freeman became the soul mate Carson had never known before. Dorothy and her husband shared the same view Carson saw from Silverledges. In the summer of 1952 their son gave them a copy of *The Sea Around Us.* The family read it aloud, marveling as it unveiled the life teeming beneath the ocean's surface. Now they welcomed its famous author who lived just around Dogfish Head.

Carson certainly did not seem imposing when Dorothy first met her. At the age of 45, Carson stood just 5′4″ and weighed barely 110 pounds, with short, wavy auburn hair swept off her forehead. Dorothy could scarcely believe that this new neighbor was both an established scientist and an author of two major books. While many people found Carson somewhat distant upon first meeting, Dorothy recognized the warmth beneath her reserve. And once she got over Carson's celebrity, she discovered their many connections. Both loved cats. Both had an elderly mother to care for. Above all, both harbored a passion for nature and especially the seashore.

Dorothy was nine years older than Rachel when they met and in some ways a daughter of the sea. She grew up on the Massachusetts coast and had come every summer to Southport since she was an infant. Still, it was Rachel, the newcomer, who introduced her to the creatures in the tide pools. They reached the shore down a steep wooden staircase off of Rachel's deck. Low tide exposed the rich varieties of shore life—gray sea squirts, starfish, mussels, crabs of all sizes and shapes, and flowers that looked like plants but were actually tiny animals. At times the two crawled, even wriggled on their stomachs, to poke under the seaweed and reach into the crevices below the tide line. At the end of the day, they brought their specimens to Rachel's cottage, where they examined them under a microscope. When they were done, Rachel returned the creatures to their home in the sea just as her mother had taught her as a child.

Dorothy was a self-trained naturalist who closely observed the life around her. As a scientist, Rachel taught her to see more systematically how climate and geography influenced species— where they lived, what they looked like, and how they survived. For her part, Dorothy provided Rachel an emotional sustenance that was missing from her life. Maria Carson's neediness left Rachel isolated in ways old friends and associates such as Rodell could ease but never alleviate. In contrast, Dorothy radiated a warmth that drew people to her. Her life was rich with friends, but none, as it turned out, proved more special than Rachel. Shortly after they first met, Rachel wrote to say, "it seems as though I had known you for years instead of weeks, for time doesn't matter when two people think and feel in the same way about so many things." Each summer meant nature walks, long talks, and the glow of Dorothy's companionship. Over the long fall, winter, and spring months, with Dorothy in Massachusetts and Rachel in Maryland, they kept their friendship alive with a steady flow of letters and phone calls as well as occasional visits.[6]

By the fall of 1953 Rachel had three friendships in place that enriched both her life and her writing. Rodell continued to manage Carson's publishing affairs and to limit the distractions that might interfere with her work. Brooks nurtured her writing and gave her the freedom to discover the approach that realized the possibilities in her subject, even if that meant long delays in producing a manuscript. And Dorothy became the person to whom Rachel unburdened herself, sharing her joys and sorrows. The professional, literary, and emotional support provided by these three sustained Carson as she struggled with her books.

For Rachel, "a writer's occupation is one of the loneliest in the world," as she warned one aspiring author. She confessed to Dorothy at one low point, "I'm still so far from the end that I feel pretty desperate about it in spite of my publisher's understanding indulgence." Now, at least, she had Dorothy, someone with "the depth of understanding to share, vicariously, the sometimes crushing burden of the creative process." A less ambitious person might simply have chosen to live off her fame and fortune.

Carson knew, however, that she was most alive when she was writing, no matter how painful it might be. As Brooks put it, writing might be misery, but "the only greater misery would be not to write at all."[7]

As with each of Carson's previous books, the shore guide evolved slowly in her mind. She first mentioned the idea for a book "on the lives of shore animals" to Rodell in 1948, even before she received a contract for *The Sea Around Us*. In December 1949, when she met Brooks, he described *Under the Sea-Wind* as "a superb job in a field where really good books are rare." It was at that meeting that Brooks proposed a book to educate the public about the life cycles of shore creatures. He encouraged her to model it after the enormously successful field guides to the birds of North America written and illustrated by Roger Tory Peterson and edited by Brooks himself.[8]

Having accepted in principle an idea in keeping with her own thinking, Carson began to elaborate what she had in mind. "My quarrel with almost all seashore books for the amateur," she wrote Brooks the following July, "is that they give him a lot of separate little capsules of information about a series of creatures, which are never firmly placed in their environment." She proposed, much as she had done in *Under the Sea-Wind*, to depict "what life may be like in terms of a fiddler crab's existence, or a barnacle's; that it should suggest unobtrusively, how the particular environmental setting (kind of shore, currents, tides, waves) determines what creatures will be found in any particular place."

Still, the concept of a practical guide remained foremost. The book would be "used by people needing information easily found, quickly read, clearly understandable." These would fall into two classes of readers, "both amateur naturalists: Vacationists whose interest in shore fauna is secondary and casual . . . [and those] who want to know where and how to collect things." She planned to "make each animal seem a living creature to show eating, mating, shelter, defense," thus placing ecology in the forefront. Finally, the book would need extensive illustrations to help readers identify what they saw.[9]

Given the prominent role of images in the book, Carson understood that the choice of an illustrator was critical to its success. Among her closest associates at Fish and Wildlife was Bob Hines, a staff artist. Indeed, Hines was the friend Carson dragged into a phone booth to share the news that the Book-of-the-Month Club was considering *The Sea Around Us* as a selection. He not only was a gifted illustrator but also shared many of Carson's values. Carson once explained to Brooks, "we want text and drawings as close to accuracy as it is humanly possible to be." Rachel also knew Hines got along well with her mother. That was vital because during the long process of research, writing, and illustrating he would be a frequent visitor to the Carson household.[10]

When not coping with her newfound fame, Carson devoted much of the next two years to research on the book. To study the ecology of the rocky shore north of Cape Cod, she had only to climb down the ladder from her front deck. There, in the tide pools she could find the starfish, sea squirts, and weeds typical of many a rocky shore. Other material required time at Woods Hole and oceanographic labs along the Atlantic coast as well as trips to the sandy shores of the Carolinas and the mangrove swamps of the Florida Keys.

Despite frustration with her slow progress, Carson loved this work, punctuated with the occasional thrill of discovery. At one point on her explorations below Charleston, South Carolina, she came upon a rare rock outcropping. Suddenly, she "felt as if I were in Maine again, for there were limpets, chitons, periwinkles, etc., of the Maine coast, yet with the blue Gulf Stream water almost washing the rocks." From that discovery she came to understand that beyond latitude and climate local conditions, such as soil and terrain, determined the ecology of an area.[11]

As always in the research for her books, Carson constructed a network of scientific experts. For this she needed to know far more than simply who was who in the world of science. She also had to ask the right questions and translate what she learned into language a general reader could comprehend. There was the big picture as well. Carson did not want her book to be a compendium of isolated facts. In time, she recognized the need to identify individual species and to locate them within the larger whole. She struggled to find the right balance. "As I'm sure you will understand," she wrote to an English scientist, "this shore book is proving the most difficult one I have written. For every writer, I suppose, there is one 'right kind' of book to write, and I think I have spent a lot of futile effort on this one fighting the subject and trying to write the wrong kind of book."[12]

Carson found the balance she sought in the Florida Keys. Before then, the "attempt to write a structureless chapter that was just one little thumbnail biography after another was driving me mad." Standing amidst the corals and mangroves, she realized that she had been looking at four different types of shores, though she soon reduced them to three: the rocky shores of New

England north of Cape Cod where the tides determined the flora and fauna, the sand beaches that stretched south from the cape into Florida where the waves more than the tides defined life, and the far southern coral and mangrove coasts shaped by the ocean currents.[13]

That revelation began to shape the book. Originally, Carson assumed that, unlike *The Sea Around Us,* what she came to call *The Edge of the Sea* would speak to a limited audience. She wrote Rodell, "Whatever that book [seashore guide] earns in royalties (including presumably a paper cover edition) will be the only income from it. There will be no foreign sales—it would be no good in England or Scandinavia. We can't sell a practical seashore guide to *The New Yorker.* No one will want to reprint chapters of it. So the subsidiary income is nil." Now she understood that what she was observing along the Atlantic shore was true for similar coastal ecologies around the world. The rocky shores of New England would mirror those of Ireland and France. No longer did she see the two books as categorically different. *Edge* would be "a sort of sequel or companion volume, the former dealing with the physical aspects of the sea, this with the biological aspects of at least part of it." Of course, that meant reworking much of the original materials on the rocky shores, "but that is part of writing," she told Brooks.[14]

Carson began to envision a completed manuscript by the summer of 1954, though she worried that Hines was falling behind with his drawings. Faced with financial pressures, he was forced to accept a number of small jobs for cash. Rather than nag or press him, she concocted a more constructive and successful strategy. She asked Brooks if he would offer Hines a substantial new advance. Brooks knew the book would easily earn back the money and that Hines's illustrations would be an important part of its success, so he readily agreed to offer the money. Never suspecting Carson's handiwork, Hines thanked Brooks: "Since author and publisher are both already well established at the top of their fields, and the artist cannot say as much for himself, it seems to me that the published results can mean much more for the artist than for anyone else. And since I am said artist, I am and will continue doing my best." Carson rewarded Hines's loyalty by ensuring that his name was prominently displayed on the

Carson and Bob Hines exploring along the Maine coast. The two were determined to have illustrations that faithfully rendered the shore creatures. (SOURCE: *Courtesy of the Rachel Carson Council.*)

book jacket. He returned the favor by providing illustrations that received widespread critical acclaim.[15]

In February 1955, Carson sent Brooks the completed manuscript of *The Edge of the Sea*. To finish the book, she "canceled all invitations—both to friends who had thoughts of visiting me and *by* friends who wanted me to do things with them!" A sense "that the end seems *never* to quite come" compromised the elation she felt at having concluded the project. She worried, too, that readers and critics would not find the new book up to the standard she set in *The Sea Around Us.*

William Shawn at *The New Yorker* allayed those fears. "She's done it again," he told Rodell in June of 1954 and offered $2500 for the chapter "Rim of Sand," with an option on the entire book. Once again, he personally worked with Carson on condensing the material. To Dorothy, she explained, "I'd rather have the book appear in *The New Yorker* than anywhere else." Advance publicity was an obvious benefit, but what truly lifted Carson's spirits was Shawn's enthusiasm. It broke a crippling sense of gloom Carson admitted to suffering as she struggled with the manuscript.[16]

Brooks argued that few books of natural history succeeded as well as *The Edge of the Sea* "in bringing to life the total environment with which they deal; in combining scientific fact with personal enthusiasm." The book was nominated for the 1955 National Book Award for nonfiction, though it lost out to *An American in Italy*, a book the reading world soon forgot. The American Association of University Women honored Carson with their Achievement Award, and the National Council of Women of the United States chose *The Edge of the Sea* as "the outstanding book of the year." Naturalist Edwin Way Teale spoke for many of her professional admirers when he praised her writing, which, despite her admitted sense of "struggle and frustration," he found "serene and fresh and strong with no residue of fatigue or stress in it—and that, in truth, is a very great accomplishment." No one could have been more enthusiastic about the book than Dorothy and Stan Freeman. "We thought we knew the sea," they said in a note of thanks for their advance copy, "but through you we discovered how slight was our knowledge. It was you who really unlocked its 'beauty and mystery' to us." She dedicated the book to them. [17]

Had *The Edge of the Sea* been Carson's first book, it probably would have inspired the same fanfare that accompanied *The Sea Around Us*. After its publication in October 1955, *Edge* quickly reached the bestseller list, but it never achieved the same level of commercial and critical success. For one thing, the Book-of-the-Month Club did not select it, even though it was serialized in *The New Yorker*. Ever the diplomat, Brooks softened Carson's disappointment by pointing out that the omission might actually boost trade sales. For another thing, Carson had the mixed fortune of competing with an unusually popular sea book, Anne Morrow Lindbergh's *Gift from the Sea*. It had come out the previous spring and remained at the top of the bestseller list for over a year. Carson liked Lindbergh's book and was relieved it was "not something sensational or trashy that holds that position." Still, *The Edge of the Sea* stayed near the top of the list for almost five months.[18]

Inevitably, most critics measured the book against its predecessor. In that regard, the judgments were mixed, though favorable. The critic for the *Sacramento Bee* captured that tone when he wrote, "this book is, perhaps, a little more fact packed and a trifle less exciting than her first." For the *New Republic* reviewer, Carson "succeeded again, although with enough weakening and diminishment" to add a tone of disappointment. That evaluation reflected in part a misreading of the book. The reviewer assumed that "Miss Carson having exhausted the heart of her subject [in *The Sea*], has been forced to move out to gather what was left around the periphery." The *Saturday Review* more properly understood the complementary nature of the two books: "Miss Carson has carved the enclosing frame for her original picture, and it is an entrancing work." The reviewer in the *San Francisco Chronicle* possibly best understood Carson's problem: "if this new book does not seem as exciting a discovery as the first, it is perhaps because Miss Carson's direct, crystal-clear style (a kind of factual poetry) is by now familiar." Familiar or not, most reviewers agreed with the *New York Times* that her new book "was equally wise and wonderful."[19]

Ornithologist Robert Cushman Murphy had described *The Sea Around Us* as an "Odyssean account of the cosmic forces at work in the sea." By contrast *The Edge of the Sea* was a more intimate book. Where Carson peered through a telescope in the ocean book, here she examined the shoreline through a microscope. She wrote about what she saw, smelled, touched, heard, and sometimes tasted and put herself on every page, sometimes in the first person. As the *Saturday Review* put it, "though Rachel Carson does not literally take the reader by the hand, she is always close by, looking into tide pools, walking along sandy beaches, or clambering over slippery weed-covered rocks." The *New York Times* noted more whimsically, "You find yourself taking an enormously friendly interest in all sorts of spiny and slime-wreathed creatures you've hitherto regarded with hearty loathing."[20]

While Carson explored the three distinct shore habitats, a single theme unified the book. Each region might have a distinctive ecology, but within all three the flora and fauna evolved to meet the conflicting demands of the places where land and sea meet. An area that most humans viewed as of little value actually teemed with life. In an environment with extremes of moisture, temperatures, storms, currents, and tides, living creatures must adapt to survive. Here, Carson observed, "life displays its enormous toughness and vitality by occupying almost every conceivable niche."[21]

Early in the book, Carson described her unexpected meeting with a small, nocturnal ghost crab "lying in a pit he had dug just above the surf, as though watching the sea and waiting." Around her was no other visible life, "just one small crab and the sea" and "no sound but the all-enveloping, primeval sounds of wind blowing over water and sands, and of waves crashing on the beach." Drawing out the import of this intimate moment, she recalled "that I knew for the first time the creature in its own world—that I understood, as never before, the essence of its being." For her, time stood still as she looked at the crab. It became "a symbol that stood for life itself." Here, Carson revealed her sense of a cosmic principle as great as any she invoked in her magisterial work on the sea. That crab suggested to her "the delicate, destructible, yet incredibly vital force that somehow holds its place amid the harsh realities of the inorganic world."[22]

The Edge of the Sea explored another theme Carson developed in her earlier writing—the interdependence of all living things. This idea reflected her affinity to the values of the recently disbanded Ecology Group. While still working on the book, Carson made interdependence the central point of a talk she gave to a meeting of the American Association for the Advancement of Science in December 1953. The majority of the scientists sitting before her, she recognized, did their work in a laboratory with specimens isolated from their natural habitats. To understand a clam, they dissected it first and thought less about where and how it lived and more about how it looked through a microscope.

Carson used a microscope as well, but for her, the seashore as much as the laboratory was the place for research. Only there could one observe the relationship of living things to the larger whole. Each fit into "the complex pattern of life. No thread is found to be complete unto itself, nor does it have meaning alone. Each is but a small part of the intricately woven design of the whole. . . ." As she looked out upon that audience of distinguished scientists, she spoke confidently about her own beliefs: "The edge of the sea is a laboratory in which nature itself is conducting experiments in the evolution of life and in a delicate balancing of the living creature within a complex system of forces, living and non-living. We have come a long way from the early days of the biology of the shore, when it was enough to find, to describe, and to name plants and animals found there." Now, Carson asserted, scientists want to know, "Why does an animal live where it does?" "What is the nature of the ties that bind it to its world?" Answers to those questions, she speculated, might lie in "the biological role played by the sea water." As she explained it, "in the sea nothing lives for itself." Each life form alters the chemical nature of the water it inhabits, creating conditions conducive to other living things. "So the present is linked with the past and the future," she concluded, "and each living thing with all that surrounds it." In that way, she made ecology the central theme of her book.[23]

Sitting by a tide pool during one of her research outings, Carson noticed some sponges—creatures simple in structure and little changed since the Paleozoic Era some 300 million years earlier. Suddenly she felt linked to an ancient past. As she studied them, a fish entered the pool. Compared to the sponges, that fish struck her as a Johnny-come-lately, with an ancestry of no more than 150 million years. And to Carson, who beheld the two "as though they were contemporaries," the thought occurred that humans were biological infants "whose ancestors had inhabited the earth so briefly that my presence was almost anachronistic." Such observations challenged the anthropocentrism with which most humans view the natural world. Even more, they evoked the sense of wonder that readers earlier discovered in *The Sea Around Us*.[24]

After the publication of *The Edge of the Sea*, Carson was adrift. What to do next? She had a contract with Harper & Brothers for a book on evolution in their World Perspectives Series. She signed it just as she began writing the shore guide, but now all she had was the contract without a clear idea about the substance of the book or what shape it might take. At that moment, another project attracted her attention—a script for an educational television series called *Omnibus*. The producers offered her wide latitude in determining the subject matter and the content of the script. They insisted only that it be about "clouds."

In 1955, television was familiar to most Americans but new to Rachel. She rarely watched and admitted to her friend Dorothy, "I should see television now and then—I don't know what these programs are like." But the prospect of writing for an audience larger than the readership of all her books combined was irresistible. She realized she had the chance "to present certain facts, or to foster certain attitudes, that I consider important in a new way."[25]

In January 1956, after a month of writing, Carson delivered a script she called "Something About the Sky." Several months of negotiations followed before she and the producers did the final editing of images and text. On March 11, 1956, "Something About the Sky" aired on national television. As the camera panned the heavens, a deep voice sounded Rachel's words. Clouds, the narrator explained, are "the writing of the wind on the sky." More to the point, he added, "they are cosmic symbols, representing an age-old process that is linked with life itself." It was pure Carson—fluent, poetic, conjuring the majestic and the mundane with ideas and images she developed in her books. The show was also a popular and critical success. Even Rachel liked it. After seeing it, she went out and bought a television for herself, a decision that especially pleased her mother, confined as she was to the house by declining health. Rachel reported to Dorothy that Maria was "happy as a cat with a saucer of cream" and contentedly watching *The Lone Ranger*.[26]

Still, for all its satisfactions, "Something About the Sky," was only a television script, and Rachel's restless mind hungered for a more challenging project. She turned her attention back to evolution and her unfulfilled contract with Harper & Brothers. "I feel I have already entered upon what Edwin Teale calls the 'delightful stage of research' for the evolution book," she confessed to Dorothy. "It is a luxury just to read, and not try to write. So you see darling, *The Edge* is becoming a little dim in my mind, with my thoughts more on the future than the past."[27]

Looking forward to the future did not mean Rachel completely escaped her past. As with the early stages of her other books, this one as yet lacked focus. Although a book on evolution had the potential to be "perhaps more important than anything else I've done," she found herself in the familiar rut of having a subject without a story. At just this anxious moment, biologist Julian Huxley published his own book on the subject, what became the widely read *Evolution in Action*. A second book on evolution seemed redundant to Rachel. She shifted her focus from evolution to ecology with a new working title, *Remembrance of the Earth*.[28]

The title was catchy, but Rachel still could not grasp "the scope of my thrust for the new book." More irksome still, the series editor, Ruth Nanda Anshen, was so overbearing and intrusive that Rachel could hardly stand to work with her. Anshen, she complained, "was so confusing that now I haven't the slightest idea what [the book] is about." As Rachel's irritation grew, even the thought of being in the same city with Anshen was too much to bear. "I shall never let her know when I'm to be in New York," she sputtered, "and will always be setting out for Antarctica if I hear she is coming [to Maryland]."[29]

As much as Carson enjoyed the research for *Remembrance of the Earth,* the theme of the book continued to elude her. With a sense of relief, she turned her energies to an article commissioned by *Woman's Home Companion* that she called "Help Your Child to Wonder." It featured her grandnephew Roger, who even helped to select the adventures she presented in the article. By describing their relationship, she offered a brief glimpse into a private world of family she jealously guarded. When Roger's mother Marjorie read the article, she burst into tears. Dorothy Freeman was so moved by seeing her friend's inner self in the writing that she regretted sharing the private Rachel with the world.[30]

Behind the mask of the disinterested writer, Carson used the article to express some of her most deeply held beliefs and feelings. Most powerfully, she argued that for a parent to guide a child "it is not so important to *know* as to *feel.*" Surprising as that assertion was for a scientist, Rachel went further: "Once the emotions have been aroused—a sense of the beautiful, the excitement of the new and the unknown, a feeling of sympathy, pity, admiration or love—then we wish for knowledge about the object of our emotional response." Not only did this piece move her friends, it gave Carson enormous personal satisfaction. Beyond sharing her passion for nature and its beauties, she included a special tribute to Dorothy in a passage about a cloudless night spent "with a friend" out on Dogfish Head. The two lay gazing up at "the misty river of the Milky Way" overwhelmed by "the millions of stars that blazed in darkness." Dorothy, too, recalled that magic moment that the two shared as "Our-night-on-the-Head."[31]

Such happy memories were a welcome escape from the growing difficulties Carson was having at home. She worried most about her mother's declining health, both physical and mental. Crippling arthritis in Maria's knees prevented her from walking. She became increasingly suspicious that Paul Brooks and Marie Rodell were displacing her in Rachel's heart. Diabetes disabled Marjorie to the point that she could no longer work or care for her son Roger. Carson realized she had to find a house big enough so that she could look after her family and still have a place to write.

The person to whom Rachel could confide her mounting anxieties was Dorothy Freeman. Despite, or more possibly because of, their common struggles with aging mothers and sick relatives, their friendship flourished. It was to Dorothy that Rachel confessed her emotional exhaustion: "Meanwhile what is needed is a near-twin of me who can do everything I do except write and let me do that!" she wrote in the spring of 1956. "It seems so silly to be spending my time as a nurse and a housemaid."[32]

Carson's normal escape to Maine over the summer of 1956 was taken up with nursing her family. "For the past three weeks," she wrote to Rodell from Southport, "I have been making daily trips to the clinic where Mamma has been having treatments for her knee and Marjorie for her shoulder, which was fast approaching the frozen state." Rodell fully sympathized with Carson's struggle. But practical person that she was, Rodell recommended practical solutions that centered on the need for domestic help. Carson knew that Rodell meant well but came to resent "her brisk competence and her 'slide-rule solutions' " that led Carson "to retreat further into myself." She simply could not move her mother into a nursing home or face her mother's displeasure at having a nurse running her house. So Carson persevered even though the family burden made writing a near impossibility.[33]

The need to improve domestic arrangements in both Maine and Maryland renewed Carson's financial worries. Rachel fretted about money for another reason, one rooted not in nurturance but in nature itself. She and Dorothy shared a dream—to preserve forever the forest tract behind their Southport homes. That place was a favored retreat "with its mossy cliffs and its unexpected open patches where only the reindeer lichen covers the rock."

With an aging mother and sick niece to care for, Rachel simply could not afford the purchase price. She told Dorothy and Stan, "if ever I wish for money—lots of it—it is when I see something like [the adjacent woods]. Just for fun, tell me what you think and let's pretend we could create a sanctuary there, where people like us could go . . . " 'and walk about and get what they need.' " Thus, she took on projects that promised large royalties without taking too much time—among them, a child's version of *The Sea Around Us* and an anthology for a series called the World of Nature. She imagined the anthology as encompassing "the whole story of Life on our earth, the geologic setting, the beginnings of life, its amazing ramifications and adaptations, [the Darwinian outlook] its relations to its physical and biological environment."[34]

A missionary zeal again fired her imagination: "On a canvas of such noble scope, what a picture one might paint! I am sure this is the right approach—the opportunity to make a true contribution to popular thinking." Equally important, the anthology might put her writing back on track. "In fact, I have dimly sensed that such an anthology could be important for *me*," she explained to Dorothy. "I'm still groping and can't say clearly what I mean but I can see at least sketchily the outlines of perhaps the rest of my 'life's work' in writing, and this seems to be a step in it. All part of a vast theme that others keep trying to state effectively, but can't. Maybe I can't either, but I keep feeling I can almost see how, and perhaps if I try long enough, I can." Whatever its possible significance, the anthology project withered away, though the "Jr. Sea" was published, with the royalties assigned to Marjorie and Roger.[35]

By early 1957 Carson's world seemed to be falling apart. The previous fall "she had someone sick on her hands" all the time and could hardly imagine being carefree ever again. In early January, a case of sniffles and aches turned into the flu. No sooner did she recover than the diabetes that afflicted Marjorie brought on a severe pneumonia. Within a month she was dead. Her loss devastated Carson, not the least because she now had young Roger to raise on her own. At the time his mother died, five-year-old Roger was also suffering from the flu. The normally lively child was bedridden for over a month. Carson, a none-too-robust woman of 49 years, now had her crippled mother and her grandnephew on her hands. When healthy, Roger had "ants in his pants," but he was often sick and unhappy.

To Carson only one solution seemed possible: she adopted Roger. He "had lost his father before he could remember him," she reasoned, "and in our small family I am the logical one to care for him and, I'm sure, the one who is closest to him." As a practical solution, she decided to build a new house in Maryland and to expand her cottage at Southport with a new dining room, a larger porch, and a room for Roger, though without the bunk beds she feared would become platforms for "fancy high-dives" off the upper level. By the end of May, family duties consumed so much of her time and energy that she realized, "I'm not a writer at present."[36]

Yet Carson lived for her research and writing. In December 1955, when she still hoped to write her evolution book, she described to Dorothy what her work meant to her: "I am taking to research like an old alcoholic to his bottle. Really, it is so stimulating, and I find my mind in a ferment of ideas." Several months later, she tried once again to explain these complicated feelings to her friend. It was if she needed to warn Dorothy that she sometimes disappeared into her work. "I think once (a long time ago) talking of *The Sea*, I tried to describe the thing that happens when one has finally established such unity with one's subject matter that the subject itself takes over and the writer becomes merely an instrument through which the real act of creation is accomplished," she confessed. "And when this mysterious something happens, one always knows it." Carson had felt that transformation when writing *The Sea Around Us* but only in short flashes when working on *The Edge of the Sea*. Now, she craved that passion again, for it was a "problem I must solve if I am ever again to be the writer I could be."[37]

Somehow, in the midst of all her personal trials, Carson found her subject. Its broad theme—"Life and the relationship of Life to the physical environment"—she intended to explore in her

Harper book, *Remembrance of the Earth*. In struggling to understand just what she wanted to say, she came to see "the whole world of science has been revolutionized by events of the past decade, or so." For Carson, a shift in thinking began with the dawn of the atomic age. Before the ominous mushroom cloud appeared over Hiroshima, "It was pleasant for me to believe, for example, that much of nature was forever beyond the tampering hand of man—he might level the forests and dam the streams, but the clouds and the rain and the wind were God's. . . ."

Now, Carson found it difficult to sustain any confidence in the enduring capacity of "life" to withstand human assaults. She discovered in herself a new skepticism toward the dominant faith of postwar America in science and human progress. "These beliefs have almost been a part of me for so long as I have thought about such things," she reflected. "To have them even vaguely threatened was so shocking that, as I have said, I shut my mind—refused to acknowledge what I couldn't help seeing. But that does no good and I have now opened my eyes and my mind." The conclusion was obvious: it was " time someone wrote of Life in the light of the truth as it now appears to us. And I think that is the book I am to write." In it, she would make the case that while in the era of space exploration "man seems likely to take into his hands . . . many of the functions of God" and "he must do so with humility rather than arrogance." She knew that this appeal for humility toward the natural world was not the book she had in mind for Harper, nor would it bear the title *Remembrance of the Earth*. In searching for what she hoped to say about "Life," Carson discovered the theme of her most important work, *Silent Spring*.[38]

Looking back on the genesis of the book, Carson credited her friend Olga Owens Huckins for leading her to the subject of pesticides. As literary editor for the *Boston Post,* Huckins in 1952 gave *The Sea Around Us* a favorable review. Carson wrote a note of thanks, and the two had corresponded ever since. In February of 1958, Huckins sent Carson a copy of a letter she published in the *Boston Herald* complaining about a state program of aerial spraying to control mosquitoes. Huckins owned land next to a bird sanctuary, and after the spraying she discovered dozens of birds in their death throes around her birdbath. While the mosquitoes remained dense and voracious, beneficial and harmless insects such as grasshoppers and bees disappeared. Huckins wanted the spraying stopped and appealed to Carson for the names of people in Washington who might help her build a case.

Carson recalled that she immediately began collecting information about pesticides. The more she collected, "the more appalled I became." Soon after, she realized she had the material for a book but wanted "to do more than merely express concern: I wanted to demonstrate that that concern was well founded." Years later, she wrote Huckins to express her gratitude. "It was not just the copy of your letter to the newspaper, but your personal letter to me that started it all." In her effort to find someone to help Huckins, Carson "realized I must write the book."[39]

Certainly, Huckins played a role; but by the time Carson received her letter, she was already collecting material about pesticides and persuaded of their danger to living things. As early as 1945 Carson had approached *Reader's Digest* about an article detailing the dangers of DDT based on research being done at a nearby Maryland laboratory. Five years later, a congressional committee held hearings into the potential dangers of DDT, though it recommended no new policy. By the mid-1950s a whole range of pollution-related issues, from the brown acrid smog over Los Angeles to the phosphate-fed algae blooms choking Lake Erie, began to receive widespread public attention. As a scientist and naturalist, Carson knew that an ecological crisis was brewing and that pesticide poisoning was a major factor. She was already involved in two separate cases that drew her attention back to the dangers of DDT and the whole class of pesticides based on chlorinated hydrocarbons.

In the fall of 1957, friends from the Washington chapter of the National Audubon Society told her about a U.S. Department of Agriculture (USDA) program to eradicate the fire ant from vast areas of southern farmlands and forests. For a nation engulfed in the McCarthy era's witch-hunts, the fire ant was the insect world's perfect counterpart to Communist subversives—unseen, dangerous, and potentially deadly. It entered the United States from Brazil during World War I and spread from a small area around Mobile, Alabama, across the Southeast. The ants built as many as thirty mounded hives per acre, some as much as five feet high. These structures were almost as strong as concrete and broke the blades off farm plows. The ants, when aroused, administered a painful bite, leaving the victim's skin swollen for days. If enough of them bit a small animal, they could kill it. All the same, southern farmers managed to coexist with this pest for over forty years. Furthermore, climate factors guaranteed that this nuisance would remain confined to the deep South, for the fire ant could not survive hard winters.

Sometime in the mid-1950s the USDA determined that the fire ant posed a serious menace to people and livestock. The USDA declared all-out war on the creatures with an extermination program that called for the spraying of some 20 to 30 million acres. Echoing USDA press releases, the *Wall Street Journal* suggested that the fire ant might "rival the boll weevil" as a threat to agriculture and had already "marched, trillions strong, across more than 20 million acres in 10 Southern states." To fight this invasion, the USDA sprayers chose as their version of an entomological atomic bomb dieldrin and heptachlor, both chlorinated hydrocarbons far more lethal than DDT and serious risks to both wildlife and beneficial insects. Yet, no insect lent itself better to limited-spot spraying than the fire ant. Most likely, USDA officials saw fire ant eradication as an opportunity to wage a public relations campaign that would promote both the department's Agricultural Research Service and the interests of its allies in the agricultural chemicals industry.[40]

This public-relations campaign quickly turned into a public-relations war. A wide range of conservation groups, sportsmen's clubs, and scientists criticized the program as both destructive and unjustified. Famed ornithologist Robert Cushman Murphy, who encouraged Carson's work, described the aerial sprayers as "trigger happy." Alabama agricultural officials investigating the sprayed areas discovered a wasteland "that reeked with the odor of decaying [wildlife]." Critics pointed out that the USDA had launched the program with no plan to measure the unintended impact of the spraying. Several southern states withdrew their support for the program, which ended as both a biological and a political fiasco. Among those whose wrath USDA officials managed to incur was Rachel Carson.[41]

Carson also became involved when a second front of the war against insects opened in the Northeast. In the 1860s, an American scientist hatched the misguided plan to import the gypsy moth to create an American silk industry. Over time, the careless loss of a few eggs let the moth escape from the lab into the surrounding trees. From there the species spread, causing recurring infestations that left large swaths of forest without leaves in the middle of summer. Eradication and control had only limited success at containing the infestations. In 1957, USDA officials designed a new strategy. They launched a massive aerial spraying campaign against the gypsy moth using a mixture of DDT and oil on three million acres in Michigan, New York, Pennsylvania, and New Jersey. It was the first time, observed J. I. Rodale, popularizer of organic gardening, that the gypsy moth program for "the highly populated northeastern area of the United States had dramatized to its populace the extent to which the chemical industry is today influencing our environment." Among those whose properties were sprayed as many as fifteen times were some prominent Long Island gardeners, including Robert Cushman Murphy; Jane Nichols, daughter of financial mogul J. P. Morgan; Archie Roosevelt, son of Teddy Roosevelt; and Marjorie Spock, sister of the world's most famous baby doctor, Benjamin Spock. When they learned that the USDA would repeat the spraying the following year, they sought a court injunction to halt the program.[42]

The Long Islanders objected to the pesticide campaign on a variety of moral, legal, and scientific grounds. Little evidence existed that aerial spraying had any more than a limited short-term effect on the gypsy moth, while the collateral damage to wildlife was extensive and the danger to humans largely unknown. The litigants also objected because indiscriminant spraying gave private citizens no way to protect themselves or their property, even when high-handed government officials behaved irresponsibly. Above all, they believed the government was conducting a badly conceived experiment that put the public at risk. A speaker at the National Audubon Society meeting complained of "something in the air that smacks of dictatorship—a type of government never heretofore associated with our American way of life."[43]

In January 1958, Carson began contacting former colleagues in the government to learn more about the nature and extent of spraying programs. She also wrote to DeWitt Wallace at *Reader's Digest,* who was about to publish an article favorable to spraying. There was, Carson cautioned, "enormous danger—both to wildlife and, more frighteningly to public health—in these growing projects for insect control by poisons, especially as widely and randomly distributed by airplanes." Wallace promised only to "weigh all the facts" before publishing the gypsy-moth article.[44]

Carson also sought help from E. B. White, a regular *New Yorker* essayist and the author of such children's classics as *Stuart Little* and *Charlotte's Web.* She wanted White to involve *The New Yorker* in the Long Island case and use his literary talents to arouse public opposition to indiscriminate pesticide use. But having already written about the dangers of pesticides, White declined to do another piece on the subject. He was still concerned, all the more so after Carson informed him of plans to spray in

Maine, where he owned a house. Rather than write an article himself, he encouraged Carson to take up the cause and promised to forward her letter to William Shawn at *The New Yorker*. That was precisely what Carson hoped would happen because by February, even before she heard from Huckins, she had made the project her own. "In the course of all this, I have made certain valuable contacts and discovered many leads to be followed up," she told Marie Rodell. She even rented a tape recorder as "a quick and easy way of assembling my poison-spray material." So far, only Roger had benefited from the gadget because he was "charmed with the sound of his own voice."[45]

Carson recognized that the materials collected by the Long Island plaintiffs contained a wealth of information on pesticides, so she asked Rodell to request copies from Marjorie Spock. Spock could not have been more pleased to learn that Rachel Carson, the great nature writer, had taken an interest in the case. She and her fellow plaintiffs were soliciting support from sympathetic public figures, and Carson was a welcome addition to their roster. Along with a stack of photocopied documents, Spock included a note in which she reminded Carson "how grim the struggle with the U.S. government and the whole chemical industry is bound to be." Spock inquired if Carson would consider serving as an expert witness for the plaintiffs.[46]

At that moment, such a commitment was more than Carson was prepared to make. Her project was just getting under way. She initially saw the "poison material" as the basis for an article, if not for *Reader's Digest* then perhaps for *Ladies' Home Journal*. To that end, she prepared a memorandum for Rodell outlining some of the "horrifying facts about what is happening through the mass application of insecticides." None of the magazines Rodell approached showed any interest in tackling the controversial material. The laboratory at *Good Housekeeping* rejected the proposed article as "something which under no circumstances should we consider. We doubt whether many of the things outlined in this letter could be substantiated."[47]

Finding the proper outlet for this new material was only one problem Carson faced. Another was more personal—how would she reconcile Dorothy to this new passion in her life? By February the project had already begun to consume much of her time and energy. Every day she corresponded with a wide network of scientists, government agencies, and concerned citizens. As the Long Island lawsuit moved forward, it commanded more of her attention. That meant fewer letters and less time for Dorothy. Rachel hoped Dorothy understood her commitment to what she was doing. "There would be no future peace for me if I kept silent," she explained. "I wish I could feel that you want me to do it." Given what this might mean to generations yet to come, Carson believed she had to push on no matter what the personal cost.[48]

Dorothy did find pesticide abuse a grim subject, in contrast to the celebration of nature that had inspired Rachel's earlier books. She particularly objected to the notion of a "poison book." Even the phrase struck her as distasteful. That was not, however, the real source of her unease. In *The Edge of the Sea*, Carson wrote about the seashore world they explored together. *Man Against the Earth,* a working title for the pesticide project, was far more technical; and Dorothy, unlike Rachel, was no scientist. Shore-birds and flowers, not chlorinated hydrocarbon molecules, interested her. Dorothy also knew how quickly Rachel could become overwhelmed when she took up a new project. Now, added to the burdens of caring for Roger and her mother, Rachel was immersed in a highly controversial subject and preoccupied with gathering information. For all those reasons, Dorothy worried that the book would be a burden on their friendship.[49]

Even more, however, Dorothy feared what the additional pressure might do to Rachel's fragile health. How would she hold up under attack from angry business executives and government bureaucrats? During the past summer in Maine, Carson had struggled with nagging problems, including a stomach disorder that turned out to be a duodenal ulcer. Dorothy suspected it had something to do with Rachel's passion for her new project. Rachel then gently reminded Dorothy "we can't lay it all to that my dear, for I was miserable much of the time in Maine last year, even when doing practically nothing."[50]

As much as Rachel wanted to allay Dorothy's concerns, she remained committed to the book. Following up on E. B. White's suggestion, she proposed to William Shawn an article on how aerial spraying of pesticides threatened the balance of nature. Shawn, she suspected, would be sympathetic. He clearly admired her writing and, having published controversial articles on segregation, nuclear radiation, and the military–industrial complex, was no stranger to controversy.

Still, the prospect of a book gave her pause. The combination of prior publishing commitments, health problems, and family responsibilities made her time precious. For a brief moment, she thought she might edit a book of essays in which she contributed a chapter that Shawn could publish as an article in *The New Yorker*. That idea quickly passed. In early April, Carson spent a day talking long distance to both "New York" and "Boston." "New York" meant Shawn, who, Carson reported to Dorothy, was "quite his usual, wonderful self" and asked her for a long two-part piece. "Boston" meant Paul Brooks, her friend and editor at Houghton Mifflin, who called to discuss details for a contract he was preparing for "the poison book." A month later, in May 1958, she overcame any final hesitation and signed it. Carson was now committed to a project that would draw on her wide network of scientific experts and demand all her scientific, literary, and emotional resources. Several questions remained: Did she have the scientific skills to master such a complex and controversial subject? Were her literary talents great enough to make the technical issues clear to the general reading public? And equally problematic, could she finish such a book before her emotional resources were completely drained?[51]

NOTES

1. Carson to Dorothy Freeman (hereafter cited as RC to DF), February 1, 1956, in Freeman, *Always, Rachel*, p. 148; see also Lear, *Rachel Carson*, pp. 213–214 and 236–37.
2. Carson to Dorothy Algire, July 7, 1946, and to Shirley Briggs, July 14, 1946, RCP-BLYU.
3. Carson to Marie Rodell, September 9, 1952, RCP-BLYU.
4. Lear, *Rachel Carson*, p. 241.
5. Brooks, *House of Life*, p. 159; Carson to Paul Brooks, June 24, 1953, and September 9, 1950, and Brooks to Carson June 30, 1953, RCP-BLYU.

6. The beginning of this relationship is reflected in Carson's letters, see Freeman, *Always, Rachel*, pp. 5–7, especially RC to DF, September 10, 1953; Lear, *Rachel Carson*, pp. 244–247 is particularly insightful about this friendship.

7. Carson to Lois Crisler, August 19, 1959, RCP-BLYU; Freeman, *Always, Rachel*, p. xxvi and Brooks to Carson, in Brooks, *House of Life*, p. 164.

8. Lear, *Rachel Carson*, p. 173; Brooks to Rodell, December, 1949, RCP-BLYU.

9. Carson to Brooks, July 28, 1950, and October 14, 1950, RCP-BLYU.

10. Carson to Brooks, April 12, 1953, RCP-BLYU; Lear, *Rachel Carson*, p. 209.

11. Carson to Brooks, April 26, 1952, RCP-BLYU; see also Brooks, *House of Life*, p. 158.

12. Carson to Dr. T. A. Stephenson, May 17, 1953, RCP-BLYU.

13. Carson to Brooks in Brooks, *House of Life*, p. 160.

14. Ibid., and Carson to Rodell, September 29, 1950, RCP-BLYU.

15. Bob Hines to Brooks, March 17, 1953, RCP-BLYU.

16. Brooks, *House of Life*, p. 164; RC to DF, June 20, 1954, and June 26, 1954, in Freeman, *Always, Rachel*, pp. 46 and 48.

17. Brooks, *House of Life*, pp. 164–165; Edwin Way Teale to Carson, April 22, 1955, RCP-BLYU; DF and Stanley Freeman to RC, October 26, 1955, in Freeman, *Always, Rachel*, p. 123.

18. RC to DF, November 20, 1955, in Freeman, *Always, Rachel*, p. 137.

19. Reviews collected in RCP-BLYU.

20. Ibid.

21. Carson, *The Edge of the Sea*, (Boston, Houghton Mifflin, 1955) p. 11.

22. Ibid., p. 14.

23. Ibid., p. 39; see also Carson, notes for paper "The Sea Frontier," delivered December 29, 1953, to American Association for the Advancement of Science, RCP-BLYU.

24. Carson, *The Edge of the Sea*, p. 109.

25. RC to DF, December 2, 1955, in Freeman, *Always, Rachel*, p. 145; Brooks, *House of Life*, p. 199.

26. Brooks, *House of Life*, p. 200. RC to DF, April 8 and 14, 1956, in Freeman, *Always, Rachel*, pp. 165, 172.

27. RC to DF, November 23, 1955, in Freeman, *Always, Rachel,* p. 138.

28. Ibid., p. 85.

29. Carson to Curtis Bok, July 12, 1956, RCP-BLYU. She complained about Ansen to Dorothy and Stan Freeman; see Lear, *Rachel Carson,* p. 287.

30. DF to RC, April 11, 1956, in Freeman, *Always, Rachel,* p. 170; "Help Your Child to Wonder" was published posthumously by Harper & Row in 1965 as *The Sense of Wonder.*

31. An excerpt from "Help Your Child to Wonder" is in Brooks, *House of Life,* pp. 202–203; DF to RC, April 11, 1950, in Freeman, *Always, Rachel,* p. 170.

32. RC to DF, February 3, 1956, in Freeman, *Always, Rachel,* p. 150.

33. Carson to Rodell, August 29, 1956, RCP-BLYU; see also Brooks, *House of Life,* p. 208, and RC to DF, December 31, 1957, in Freeman, *Always, Rachel,* p. 242.

34. Brooks, *House of Life,* p. 214. Carson to Rodell, October 2, 1956, RCP-BLYU; RC to DF September 23, 1956, RC to DF and Stan Freeman, October 7, 1956, and RC to DF, December 8, 1956, in Freeman, *Always, Rachel,* pp. 190, 194, and p. 204.

35. Ibid.

36. Brooks, *House of Life,* pp. 215–216.

37. RC to DF, December 2, 1955, and February 1, 1956, in Freeman, *Always, Rachel,* pp. 145, 148.

38. Ibid., February 1, 1958, pp. 248–249.

39. Brooks, *House of Life,* pp. 233–235, helped establish the idea that Huckins had inspired the book; Lear, *Rachel Carson,* pp. 314–315, p. 545 n. 7, p. 546 n.14, by a more careful reconstruction of the process, discounts that version.

40. *Wall Street Journal,* January 28, 1958; see also Dunlap, *DDT,* pp. 89–90.

41. Russell, *War and Nature,* pp. 213–214.

42. Ibid.

43. Ibid., p. 218.

44. Carson to DeWitt Wallace, January 27, 1958, and Walter Mahoney to Carson, January 30, 1958, RCP-BLYU.

45. Carson to Rodell, February 2, 1958, RCP-BLYU; Lear, *Rachel Carson*, pp. 316–317; see also RC to DF, February 8, 1958, in Freeman, *Always, Rachel*, p. 251.

46. Marjorie Spock to Carson, February 5, 1958, RCP-BLYU.

47. Brooks, *House of Life*, pp. 236–238.

48. RC to DF, February 1, 1958, in Freeman, *Always, Rachel*, pp. 248–249.

49. Ibid., May 2, 1958, p. 256.

50. Ibid., June 28, 1958, pp. 258–259.

51. Ibid., April 2, 1958, p. 252.

· *Four* ·

WINTER

The Poison Book and the
Dark Season of Vindication

WHEN SHE DECIDED TO WRITE WHAT DOROTHY FREEMAN called the "poison book," Rachel Carson committed herself to a crusade, driven on by a sense of moral outrage. She believed that the arrogance of humankind created a deadly irony: in their determination to control nature, human beings posed a growing threat to all life on earth, including their own. Despite her forebodings, Carson never intended to let "the ugly facts dominate" the book since "the beauty of the living world I was trying to save has always been uppermost in my mind." But so too was a profound "anger at the senseless, brutish things that were being done."[1]

To promote her cause, Carson set out to build a case against those who made wanton use of dangerous chemicals. Such a case required more than gathering evidence. It required finding the *right* kind of evidence. She undertook her research as a partisan, not a neutral observer. In that sense, Carson violated the canons of scientific objectivity. Those who supported the cause were allies; those opposed, enemies. It was "us" against "them" and "our side" and "their side." She could assure a friend, as perhaps she was assuring herself, "The right people are so glad I'm doing the job" and spoke approvingly of one scientist as "so *militantly* on the *right* side." Carson banished all doubts about the righteousness of her crusade. She was out to win.[2]

Carson understood all too well that the United States in the 1950s was not hospitable to crusades against powerful interests, whether in government or in business. A post–World War II consensus dominated American society. At its core lay a profound anti-Communism that meant both containing Soviet expansion abroad and fighting subversives at home. The consensus encouraged social and political conformity, respect for governmental and community authority, uncritical patriotism, religious faith, and a commitment to a vague notion of an American way of life defined by prosperity, material comfort, and a secure home. A person did not have to be a Communist to come under suspicion as a subversive. One had only to dissent against commonly accepted values, as Carson intended to do, to be considered disloyal.

Carson's writing did not simply evoke the dangers of science run amok. A reverence for the mysteries of nature and the intricate web of life linked her to the romantics. With their idealism, passion, and fascination with the unknown world, romantics often flaunt convention. Empirical science has routinely dismissed their sensibility as "shrill," "irrational," and "emotional." These same terms distinguished the masculine from the feminine. In that way, Carson's critics could use both her passion and her gender to discredit her ideas.

Certainly, Carson's life hardly conformed to then current conventions of gender and family. In the 1950s, the ideal American woman was a suburban housewife dedicated to the career of her husband, the care of her children, and the consumption of the material bounty made available by what were generally regarded as benevolent corporations. Carson, by contrast, was an unmarried career woman who headed a family for which she, not a father or husband, had always provided. The child she adopted had been born out of wedlock and was not even her own. More problematic still, her most intimate relationship was with another woman.

Carson also resisted the consumerism of the 1950s by placing spiritual values ahead of material ones. In an age when conspicuous consumption made "keeping up with the Joneses" a requirement of life in suburbia, she was at best a reluctant consumer who treated her car, her record player, and her television as self-indulgent luxuries rather than essential commodities. Her anti-materialism, like her reverence for nature, was a value she incorporated from the stern Presbyterianism of her mother's family. That Protestant ethic contradicted the consumer ethos promoted by modern advertising. The qualities that distinguished Carson from the female stereotypes of the day endeared her to friends and associates. Those same qualities would have been ammunition for her enemies had they known more about her, but Carson never sought celebrity. She guarded her privacy so carefully that few people had any sense at all of her personal life.

Rachel knew she faced powerful opposition. A network of government scientists and bureaucrats, chemical companies, trade associations, and corporate scientists made up the "enemy" in what began as a battle between David and Goliath. Chemical and pharmaceutical companies had acquired great prestige from the vital role they played during World War II. Army medics used newly discovered penicillin to treat soldiers with infected wounds and sexually transmitted diseases. Air Crops bombardiers dropped tons of fast-burning jellied gasoline known as "napalm" to force enemy troops out of fortified positions. Dichlorodiphenyl-trichloroethane (DDT) proved invaluable in controlling the common disease-bearing insects such as mosquitoes and lice that

Dichlorodiphenyltrichloroethane (DDT). The molecular structure of DDT makes it relatively insoluble in water and, hence, persistent in the environment.

had long been a scourge of the world's armies. Most scientists were convinced that such pesticides posed no danger to humans and little threat to domestic animals or wildlife.

Carson watched uneasily in the 1950s as the Cold War offered new incentives to expand the nation's chemical arsenal, whether in the form of weapons of war or products to protect American agriculture. The military tested a wide range of chemical weapons that, unlike an atom bomb, could neutralize an enemy without widespread physical destruction. Scientists developed new classes of potent insecticides, herbicides, and fungicides to eliminate pests. Government entomologists and chemical company publicists freely employed metaphors that compared insects and Communists. At Columbia University in 1946, former British prime minister Winston Churchill suggested that Communists should study termites in order to see what their future had in store. Unintentionally clarifying the threatening metaphor, the president of the American Economic Entomologists entitled his 1947 speech "Totalitarian Insects."[3]

Though Carson did not share this view, most Americans accepted the idea that chemical weapons were vital to survival in the Cold War battle with Communism. These products would make American weapons more deadly and American agriculture more productive. And just as Senator Joe McCarthy, Federal Bureau of Investigation (FBI) director J. Edgar Hoover, and other Red hunters wanted to protect America from internal subversion, chemical companies promised to ward off the growing menace of insect infestation. After "hordes" of Chinese soldiers poured into Korea in 1951, government entomologists began to talk in similar terms about the "insect hordes" that threatened

SAVE THAT COTTON!

During periods of national emergency cotton is vital. It goes into uniforms, gun covers, pup tents, duffle bags and a thousand other items essential to the well equipped soldier. Leave it to the boll weevil and there would not be nearly enough cotton. But cotton we must have in unbelievable quantities and that means death to the weevil.

Benzene Hexachloride is the lethal chemical in most of the dust and spray insecticides which today are destroying these pests and saving cotton for the big defense job.

Whether you have weevil trouble or not, there are many ways in which your own daily life is made safer and more comfortable with chemicals from Tennessee . . . and industry serving all industry.

TENNESSEE
PRODUCTS & CHEMICAL
Corporation
NASHVILLE, TENNESSEE

PRODUCERS OF: FUELS • METALLURGICAL PRODUCTS • TINSULATE BUILDING PRODUCTS • AROMATIC CHEMICALS • WOOD CHEMICALS • AGRICULTURAL CHEMICALS

Agricultural chemical makers used Cold War themes to promote the sale of pesticides such as DDT. Many ads evoked the image of a nation at war with hordes of subversive insects. (SOURCE: Agricultural Chemicals. This image appears on p. 59 in Lytle, America's Uncivil Wars.*)*

America's food supply. One U.S. Department of Agriculture (USDA) official warned Congress in 1951 that "no enemy could disrupt our food supplies so quickly as the *hordes* of insects within our borders" (italics mine). The availability of chemicals such as DDT allowed one official to promise total victory in what was a total war. "I give an unqualified yes to the question, Can insects be eradicated?" he asserted. The reality that all-out war on insects might be dangerous and that some insects might be beneficial or even essential to life seems to have escaped this entomological cold warrior.[4]

Throughout the 1950s government support for all manner of chemical weapons against both human and insect enemies grew rapidly. So did the close ties between the chemical company executives and the government officials who supported each other's interests with enthusiasm. Such advocacy for chemicals sometimes had unintended consequences. The effort to find disabling nerve-gas agents led some people to question the military's zeal for weapons that seemed to favor the well-being of buildings over people. In an effort to find a "truth serum," the Central Intelligence Agency (CIA) conducted widespread experiments with lysergic acid diethylamide (LSD), often on unwitting subjects who sometimes felt their sanity was slipping away without knowing why. (It was not long after that that the hallucinogenic genie escaped from the laboratory into private hands.) Corporate advertising and government propaganda deflected criticism from those missteps while promoting public support for chemical weapons and wider use of agricultural chemicals. Most Americans were reassured because the DuPont Corporation promised them nothing less than "better living through chemistry."[5]

The strength of this government–industry alliance became clearer to Carson when the judge issued a verdict in the Long Island lawsuit. That case finally came to trial in the winter of 1958. Marjorie Spock and her fellow plaintiffs sought to win an injunction against aerial spraying of the gypsy moth by marshaling the testimony of expert witnesses. Ornithologist Robert Cushman Murphy described the extensive damage spraying had done to the marsh near his home. John George, a research biologist, confirmed Murphy's contention that DDT killed fish and birds. Dr. Malcolm Hargraves, a hematologist from the prestigious Mayo Clinic, cited data linking DDT to lymphoma, leukemia, Hodgson's, and other blood diseases.

To refute that line of argument, the defendants called on their own experts. Wayland Hayes from the U.S. Public Health Service pointed to experiments on convicts who consumed large doses of DDT and health statistics of workers in plants where the chemical was made. Neither group had shown harmful effects from their exposure, and Hayes thus concluded that DDT was safe. The judge ruled that the plaintiffs had failed to establish DDT as a danger. Since he concluded that the government's evidence was more convincing, he ruled "mass spraying has a reasonable relation to the public objective of combating the evil of the gypsy moth and thus is within the proper exercise of the police power by designated public officials." The disgruntled plaintiffs took their lawsuit all the way to the Supreme Court without winning the injunction they sought. Nor did their efforts shake public confidence in the safety of pesticides and aerial spraying.[6]

When she learned about the verdict, Carson understood not only the power of the "enemy" but also the need for an airtight case to win her battle in the larger court of public opinion. There were signs, all the same, of hope that she would not stand alone. Lonely voices challenging the military–industrial ethos of the day were gaining a wider audience. In 1958, John Kenneth Galbraith, an author Carson admired, published *The Affluent Society*, a critique of the cult of private consumption. The book became a major bestseller, even though the author was an academic economist. Like Carson, Galbraith was a gifted writer who could make complex ideas clear to general readers. He understood that the 1950s were not receptive to unorthodox thinking: "These are the days when men of all social disciplines and political faiths seek the comfortable and the accepted; when the man of controversy is looked upon as a disturbing influence; when originality is taken to be a sign of instability; and when, in minor modification of the scriptural parable, the bland lead the bland." More than any particular political ideology, it was the suffocating "conventional wisdom" he sought to upend.[7]

Galbraith believed that conventional wisdom created inertia and resistance to change among people of all political persuasions. Americans clung to the idea that "the increased production of goods is . . . a basic measure of social achievement." Indeed, that was a pillar upon which the current consensus rested. Galbraith argued that such productivity was not enough. What was needed was "social balance." Increases in the production of consumer goods in the private sector required a comparable increase in the services rendered to the public sector. What good were shiny new cars if roads, parking places, and traffic regulation were inadequate? He singled out Los Angeles, where failure to regulate exhaust emissions resulted in "the agony of the city without useable air."[8]

In his unabashedly liberal agenda, Galbraith called for wider opportunity for all citizens through improved schools, a renewed commitment to public service, and the emergence of a more critical public able to resist "the want creating power [to wit, advertising] which is essential to the modern economy." He proposed to achieve those ends by reinvigorating the activist state, one that served its citizens by addressing such pressing problems as poverty, inadequate schools, and disease. Galbraith did not stop there, however, turning his withering criticism on fundamental assumptions of the consensus. Few people in the 1950s, for example, questioned massive defense spending on nuclear weapons. Galbraith suggested that while hydrogen bombs might improve national security in the short term, in the long term nuclear weapons increased the risk of annihilation. Such ideas ran against the grain of the "conventional wisdom," but Carson warmly embraced them. And the success of *The Affluent Society* suggested that she, too, might have a more receptive audience by the late 1950s.

Carson found additional support from an unexpected source—President Dwight D. Eisenhower. As he neared the end of his second term in 1961, Eisenhower worried that an "immense military establishment and a large arms industry" had gained disproportionate influence in American society. In his farewell address, he urged the nation to "guard against the acquisition of unwarranted influence, whether sought or unsought, by the military–industrial complex." His solution lay with "an alert and knowledgeable citizenry" to ensure "that security and liberty may prosper together." Carson similarly saw a threat in the alliance of officials in the USDA, in state and county agricultural agencies, in the chemical corporations, and in large industrial farms popularly known as agribusinesses that waged the agricultural cold war. Like Eisenhower, she feared "public policy could itself become the captive of a scientific–technological elite." Only with the proper information, Carson believed, could the public make informed choices about how agricultural chemicals and pesticides affected their lives. She intended to give them that information.[9]

A series of public health crises prepared Americans for Carson's message. After the Soviet Union exploded an atom bomb in 1949, President Harry Truman ordered the development of the even more powerful hydrogen bomb. Americans have lived ever since with the threat of nuclear holocaust as each of the rival superpowers rapidly added to its stockpile of superweapons. Students ducked and covered under their desks as regular air-raid drills reminded them that each day might be their last, while the news media reported tests of ever more powerful nuclear weapons with enough power to incinerate the planet. Most weapons tests were conducted in remote areas far from crowded cities and suburbs. But in 1954 the crew of the *Lucky Dragon*, a Japanese fishing boat, accidentally sailed into a cloud of radioactive fallout from an American hydrogen bomb tested on a Pacific atoll. Many of the crew fell ill or died. Japanese consumers stopped eating fish for fear that fallout was contaminating their food. The incident led scientists to determine that aboveground nuclear testing subjected the world to an increased risk of radiation contamination.[10]

The primary culprit in spreading radiation was strontium-90. The isotope had an affinity for calcium. As cows ate grass dusted by nuclear fallout, the strontium appeared in their milk and then in the bones of children who drank it. Once lodged in the bones of children, it increased the risk of cancer. For a nation in the midst of a baby boom, this was terrifying news. Both Eisenhower and Adlai Stevenson raised the fallout issue in the 1956 presidential campaign. Concerned scientists, led by biologist Barry Commoner, began to educate the public and the scientific community about the danger of nuclear fallout.

Calls to ban aboveground nuclear tests became more vocal. In 1961, some 25,000 Americans staged the largest peace demonstration since World War II. The recently formed SANE (the Committee for a Sane Nuclear Policy) warned against the threat of "extermination without representation." Meanwhile, the new president, John F. Kennedy, renewed underground nuclear tests to placate Cold War hawks but also challenged the Soviet Union to engage in a "peace race" rather than an "arms race." The following November, a group of normally apolitical Washington, D.C., housewives organized the Women's Strike for Peace to call for "An End to the Arms Race, Not the Human Race."[11]

Fallout and nuclear weapons were not the sole source of public concern. The danger of agricultural chemicals raised controversy as well. Just before Thanksgiving in 1959, the Food and Drug Administration (FDA) suddenly issued a ban on cranberries containing the herbicide aminotriazole. The compound was registered with the USDA but not for use on crops. Oregon cranberry farmers, ignoring USDA guidelines, sprayed their bogs before the 1957 harvest, only to discover that a third of the crop showed traces of aminotriazole. Even before the farmers harvested the 1958 crop, tests showed that the compound could cause cancer.

Despite efforts to protect the public, growers had already shipped tainted berries. Not until November 1959 did Health, Education, and Welfare (HEW) Secretary Arthur Flemming announce a ban on cranberries harvested in 1957, 1958, and 1959. The idea that officials allowed tainted products into the food supply stunned consumers. Even worse, aminotriazole was widely used to control weeds among other fruit and grain crops. Here was clear proof of the need for more rigorous government regulation of toxic chemicals. Day by day Carson followed the aminotriazole story and especially the fortitude shown by HEW Secretary Flemming in the face of hostile industry reaction to the ban.[12]

In the summer of 1962, an incident involving the flu medicine thalidomide raised questions about the responsibility of both chemical companies and the government agencies that regulated them. A pharmaceutical company applied to market thalidomide in the United States. Dr. Frances Kelsey of the FDA insisted that the agency needed more data on the safety of the drug before approving it for general sale. Despite intense pressure from the drug company, Kelsey held firm. Researchers soon established a direct link between use of thalidomide in the first trimester of pregnancy and horrifying birth defects. In Germany and Australia, where the drug was available, babies were born without arms, with hands connected directly to shoulders, with brain damage, as well as with a range of other mental and physical disabilities. This tragedy affected as many as 8000 babies in forty-six countries. When that evidence came to light, newspapers, TV and radio commentators, and leading citizens across the nation declared Kelsey a public hero for sparing the United States a similar tragedy.[13]

Carson watched all these events with mounting concern as she collected the evidence for her "poison book," with the working title *Man Against the Earth*. Her three earlier books already demonstrated her capacity for research, her ability to synthesize complex and sometimes contradictory data, and her talent for turning even tedious information into elegant prose. Yet, a younger and healthier Carson had written those books. In order to finish this new one, she would battle her failing body as well as the political and economic interests that sought to silence her.

Carson originally contracted to finish by late 1959 what she envisioned as a short book. That schedule was unrealistically optimistic, and remembering the slow pace of her earlier books, she probably knew it. Her subject was infinitely complex, the documentation was scattered in laboratories and official files, and many potentially useful sources had no reason to cooperate. Carson soon came to realize just how formidable and time-consuming her task was. Determination to offer her enemies no avenue of attack dictated that her research be thorough and her data unimpeachable.

Despite her intention to work quickly, she could never overcome the limitations imposed by her family and her own mounting health problems. In the fall of 1958, as Carson was deep into her research, her mother suffered a major stroke, soon complicated by a serious case of pneumonia. She knew her mother could not long survive. When it was practical, she brought her home from the hospital so that Maria could spend her final days in familiar surroundings. Every day Rachel clothed, fed, and washed her mother, sometimes without help from a nurse or caregiver. Whatever emotional and physical energy she could muster went to bringing Maria comfort over her last days.

Maria Carson died on December 2, 1958, with Rachel at her bedside holding her hand. In the sad weeks that followed, Carson found comfort in Dorothy Freeman and Marjorie Spock, who stayed in close contact with their grieving friend. Rachel's loss was compounded by the need to console Roger, who had been a special object of Maria's affection. Now that his grandmother was gone, Rachel worried about Roger because Maria's death "is obviously recalling to him all the memories of the loss of his mother" less than two years earlier.[14]

She buried her mother's ashes in a cemetery plot in Silver Spring, Maryland, next to one she had purchased for herself. That was the best she could do to preserve a trace of the powerful bond that had united mother and daughter. The obituary she wrote for Maria said simply, "Maria Carson had a life long interest in nature and land conservation, which was transmitted to her daughter Rachel, whose book, *The Sea Around Us*, was a bestseller in 1952." In her loss, Carson also found freedom. She had been readying herself for this event for years. Now, with the burden of her mother's care lifted, Carson devoted herself to *Man Against the Earth*. Her mother had helped to prepare her for that. From her, Carson learned what the great naturalist Albert Schweitzer called a "reverence for life." Maria also bequeathed to her daughter, as Rachel understood, a fierce determination to fight "against anything she believed wrong, as in our present Crusade!"[15]

The crusade encouraged Carson to reach out to friends old and new. Women she knew in service clubs and government agencies provided much-needed information and support. She also built a new network of experts to guide her in her research. Clarence Cottam was one person from whom she received both encouragement and vital information. She first met the noted wildlife biologist through her work at the Fish and Wildlife Service. In 1945, her former boss Elmer Higgins and Cottam were researching the impact of DDT on fish and wildlife. Carson, who edited their controversial reports, became convinced of the importance of their work. It was then she had proposed to *Reader's Digest* an article based on their findings. Even though the *Digest* turned her down, Cottam never lost interest in the issue. He went on to direct the Welder Wildlife Foundation in Texas and from that position became one of the most outspoken critics of the USDA's fire ant–eradication program.[16]

In November of 1958, shortly before Maria Carson died, the National Wildlife Federation asked Carson to speak about the danger of insecticides to human health. Clarence Cottam chaired the panel. The need to care for her mother prevented her from accepting, but the occasion allowed Carson to renew their friendship. She confessed to Cottam that her current project had not sufficiently matured for her to make her findings public. The facts she uncovered "are pretty terrific and should not be revealed until they can form part of the total book." More than that, "the whole thing is so explosive and the pressures from the other side so powerful and enormous, that I feel it far wiser to keep my council insofar as I can until I am ready to launch my attack as a whole." Carson's concerns were confirmed when Marjorie Spock informed her that toward the end of the meeting several USDA officials overheard someone announce that Carson was writing a book on pesticides. From then on, Carson found her access to government research restricted. Cottam sensed the project would test her and warned that "because of the controversy I doubt that [the book] will ever be a bestseller regardless of how beautifully it is written and how valuable it may become."[17]

The meeting provided Carson with another valuable contact through whom she developed a major, and even more controversial, theme for her book. When she declined the invitation to participate, the Wildlife Federation replaced her with Dr. Malcolm Hargraves, the hematologist from the Mayo Clinic who had testified in the Long Island trial. Despite the skepticism of his colleagues at the clinic, Hargraves was persuaded that a link existed between pesticide spraying and leukemia. He told Carson about a research project in Memphis, Tennessee, where the USDA had undertaken an eradication program using the powerful pesticide dieldrin on both houses and yards. Now, Carson had an opportunity to meet him in person and learn more about his work.

Hargraves confirmed the cancer–pesticide connection that he and toxicologist Morton Biskind had presented in the Long Island trial. Where Hargraves focused on carcinogens, or cancer-causing agents, and blood disorders such as leukemia, Biskind researched the impact of industrial pollutants on the human enzyme system. Both came to the same conclusion: pesticides and other synthetic chemicals were carcinogenic. Carson, who read Biskind's work avidly, was convinced they were right. Their research led her to Dr. Wilhelm Hueper, a specialist on environmental cancers at the National Cancer Institute. Well before most scientists saw the connection, Hueper linked pesticides and cancer and labeled DDT a carcinogen.[18]

This line of inquiry presented Carson with a series of problems. For one, Hargraves, Biskind, and Hueper worked on the cutting edge of their fields. Many well-regarded scientists challenged their findings and, even more, criticized their willingness to speak out in public without what critics regarded as conclusive evidence. These researchers believed, however, that, given the health issues involved, the public had a right to know. Hargraves, Biskind, and Hueper collected hundreds of case histories supporting their theories; and while they found no smoking gun, they did have a mountain of circumstantial evidence. They, and through them Carson, believed that the evidence was overwhelming, and because they were willing to say so publicly as expert witnesses, the three ran afoul of the medical establishment and the chemical industry. Carson knew that the debate over their research only added to the controversy likely to explode when she published her book.

Yet another problem confronted Carson, this one involving the complexity and technical nature of the evidence. Amassing the data and reducing it to terms that general readers could understand slowed her work. Having promised Brooks to deliver the book by the end of 1959, she was hopelessly behind schedule. The cancer research, though promising, was sure to cause even more delays. Carson knew that in order to speed up her project she needed help, so she advertised for a secretary and research assistant. Among those who replied was Jeanne Davis, a slender, dark-haired mother in her early forties. Secretarial skills were but one of her many qualifications. Davis had conducted medical research and edited journals while her husband completed his surgical residency at Harvard. As a result, she knew how to find the materials Carson needed to build her case about the cancer risks from pesticides. Equally important, Davis was temperamentally matched to her new boss. She was never ruffled, kept Carson's files meticulously organized, and believed in the work Carson was doing. Carson sometimes paired her assistant with her old friend Dorothy Algire, whom Carson had met while doing research at Woods Hole and who currently worked at the National Institutes of Health (NIH). Like Carson, Algire had a master's degree in biology. Her position at the NIH gave her access to documents not readily available to other researchers.[19]

With this new team in place, Carson began to make progress. Even then, she had no hope of meeting her deadline. "You are the most patient of editors," she told the forbearing Paul Brooks. "I guess all that sustains me is a serene inner conviction that when, at last the book is done, it is going to be built on an unshakable foundation." Having missed her first deadline, Carson set February 1960 as a possible target, though she knew that was more than a little optimistic.[20]

All the same, *Man Against the Earth* was beginning to take shape. Carson had already drafted two chapters on cancer and other chapters on groundwater, insect resistance, soil contamination, and birds and other wildlife. A more technical chapter on cell biology and genetic mutation was under way, complicated somewhat by Marie Rodell's sage advice to "avoid the scientific word" whenever possible.

To Carson's delight, the American Academy of Arts and Sciences honored Hueper for his distinguished contributions in cancer research. Carson considered it "overdue recognition" and realized with wry understatement "the chemical companies won't be happy." Better yet, William Shawn at *The New Yorker* called on Christmas Eve to tell her that he found her material on insect resistance to pesticides exciting. It was work of great importance, he added. Shawn's praise was the tonic Carson needed, for as she told Dorothy, "I'm confident now about all of it."[21] Dorothy shared Shawn's enthusiasm. She had noticed "especially this fall you seemed to be deriving such satisfaction from your work," and now she regretted that "I did not encourage you from the start." Dorothy worried only about the glare of publicity disrupting her friend's life. "Will you be dragged to hearings, etc.?" she wondered.[22]

Just as all the pieces were falling into place, Carson's body betrayed her. The flare-up of a recurring duodenal ulcer forced her into bed for several weeks. By January 21 Carson reported that she had recovered enough strength "to sit up in a chair and eat my lunch." Her doctor even allowed her to nibble on stewed chicken for dinner. Meanwhile, the work mounted, compounded by pressure from Oxford University Press to deliver the material for a revised edition of *The Sea Around Us*. Carson soldiered on, though on many days she could barely leave her bed and wondered "whether the Author even exists anymore."

By March she managed to send Brooks the two cancer chapters already vetted by a researcher at Sloan-Kettering Institute in New York. She heaved a sigh of relief, feeling that "by far the most difficult part is done and the rest should roll along rather speedily." There was a complication however. She revealed to Brooks, "I am entering the hospital soon—probably this week—for surgery that I hope will not be too complicated but that can't be known at this moment."[23]

Given her history of cysts and breast tumors, Carson could not avoid surgery no matter how much it disrupted her writing schedule. As it turned out, the operation proved far more complicated than she had anticipated. Her doctor found several cysts, one of them sufficiently suspicious to require a radical mastectomy. Recovery was slow and painful, eased only by her doctor's assurance that the pathology report indicated no malignancy, only "a condition bordering on malignancy." He recommended no further treatment. Carson breathed another sigh, assuming "there need be no apprehension for the future." Within two weeks she was well enough to ask Dorothy to visit and to spend a few hours a day on her correspondence. Serious work proved too taxing, though her two assistants helped her make some progress. Jeanne Davis came three times a week to take dictation, transcribe tapes, and manage household affairs. Equally important, Davis became a friend and companion who sustained Carson through these trying months.[24]

Bette Haney was another assistant Carson hired to do some of the research she had neither the energy nor the opportunity to do on her own. A bright former biology major with experience in government, Haney knew how to find the precise information Carson sought. It wasn't always easy, and cautious government officials sometimes made Haney's task more difficult. In the early spring of 1960, Haney arranged an interview with Justus Ward, the head of the Agricultural Research Service's regulatory division. Partway through their conversation about agricultural chemicals, Ward became suspicious that Haney was not a government employee as he had thought. For whom did she work, he wanted to know? When he learned that Haney was Carson's research assistant, he ended the interview on the spot. In telling Carson about her interview, Haney described how nervous Ward seemed, to which Carson replied "He should be." In the end, Carson used her extensive network to get the information from another source.[25]

Trouble threatened from another direction. Carson knew William Longgood from his sympathetic reporting of the Long Island spraying case. That experience led him to write *The Poisons in Your Food,* a stinging critique of the government's failure to protect the food supply from chemical contaminants. Eager to discredit both Longgood and the book, chemical companies accused him of gross inaccuracies. While Carson had not read *The Poisons in Your Food* and could not testify to its accuracy, she learned from Longgood's experience that "one cannot accept any statement from whatever source as truth, until one takes the time to trace it to its original source." That, she added, "is one reason I am being so slow": she feared errors might creep into her book.[26]

By the summer of 1960, Carson was well enough to travel to Maine. It was a time for recovery and for Dorothy but not for writing. After several rejuvenating months poking around in the tide pools and meandering through the woods near her cottage, she left Southport in early September, regretting she had "so little time to enjoy the place" even though she "never loved it more." Dorothy faced her own ordeals as her aged mother and her husband Stan both struggled with serious illnesses. Rachel could take comfort that their time together, however brief, "closed that long gap of years" since she had last been in Dorothy's house at Southport. The death of Dorothy's mother in October only reminded Carson of her own uncertain future. She confessed that her "sense of urgency grows to press on with things I need to say—things that may be less important than I think, but to me at least it is necessary that they be said."[27]

Despite her health problems and writing commitments, Carson made time in the fall of 1960 to support John F. Kennedy's election campaign. She had always been a progressive, New Deal–style Democrat who believed in an activist government. She was, of course, especially critical of the Eisenhower administration's lax policy on chemical spraying programs and hoped Kennedy would take a more enlightened approach. The previous June she had been appointed to the National Resources Committee of the Democratic Advisory Council. She urged the Democrats to take up such issues as pollution control, radioactive contamination, protection of vital habitats, and wilderness preservation. The Democrats eventually addressed all those issues. In October, Carson's efforts led to her appointment to the Women's Committee for New Frontiers, a label inspired by Kennedy's pledge to get the country moving in new directions. At a lunch meeting at the home of the senator and his wife, Jackie, Carson mingled with such distinguished guests as Eleanor Roosevelt and former New Deal labor secretary Frances Perkins. She left the luncheon encouraged by the high quality of the ideas discussed and by the people she met.[28]

The Women's Committee meeting proved a rare distraction over the fall. Whatever energy she could muster went into finishing the book. Paul Brooks gently inquired if they might consider publication over the summer of 1961. To that Carson gave a re-sounding no but sent along a revised chapter on cancer. Rodell and Brooks met in November to go over the material they had received. Rodell assured Carson "that when it was good, it was very, very good, and when it is not, it is difficult." Urging her client to simplify whenever possible, Rodell warned that "references to and comments about the chemical companies need to be phrased very carefully" so as to avoid at all costs "giving anyone the opportunity to yell 'crank.'"[29]

Then, there was the problem of the title. As with each of her previous books, Carson grew dissatisfied with her working title as the book took shape. Brooks had suggested *Silent Spring* as the title for her chapter on birds. Rodell, who particularly liked that chapter, wondered "if 'Silent Spring' mightn't make a title for the whole book?" Carson was hesitant until the following August, when Rodell suggested two lines from Keats for the motto page:

> The sedge is withered from the lake,
> And no birds sing.

With that, *Silent Spring* was born.[30]

But would Carson be silenced before she could finish it? In late November 1960, she discovered a hard swelling on her left side. After considerable hemming and hawing, her doctor recommended X-ray treatments. Immediately after, she came down with a flu that left her bedridden for two weeks. Only when she arranged to go for a consultation with Dr. George Crile, a Cleveland oncologist and old friend, did her doctor confirm that her surgeon had misled her after her previous operation. The tumor he had removed was malignant. Braving a fierce winter storm, Carson went to see Crile, who confirmed the diagnosis. She was grateful that he had "enough respect for my mentality and emotional stability to discuss all this frankly." As with her research, so with her health: she had "a great deal more peace of mind when I feel I know the facts, even though I might wish they were different." She realized that treatment would cause a "pretty serious diversion of time and capacity for work because there are some side effects." Nonetheless, she planned to "work hard and productively" because more than ever "I am eager to get the book done."[31]

What ensued was a train of woes worthy of Job. A minor bladder infection gave way to blood poisoning. Antibiotics brought on a painful phlebitis that left her knee and ankle severely swollen. Unable to stand, much less walk, she was confined to a wheelchair. By February 12 the combination of disorders became so severe she was transferred by ambulance to a Washington hospital. Her doctor then diagnosed acute infectious arthritis in her joints. Even when she was well enough to go home, she returned to the hospital each day until the end of March for radiation treatments. For over a month, writing was impossible. She had earlier said to Dorothy, about the book, "I feel I can do it—if nothing happens." Now, she lost not only her capacity to write but also "any creative feeling or desire." She came to wonder if "I will walk on my beach this summer or even sit under the spruce." Still, *Silent Spring* remained very much on her mind: "About the only good thing I can see in all this experience is that the long time away from close contact with the book may have given me the broader perspective which I've always struggled for but felt I was not achieving."[32]

With the arrival of spring and the end of her radiation treatments, Carson's spirits revived as her health improved. She even began to drive on her own for short distances. By late June, she left for Maine and a quiet summer that allowed time to make progress on the book. She was "over the hump," she felt, "with just two chapters and final revisions" left to do. Her ideas had sufficiently matured that she felt confident she could make some of them public. When Brooks asked for ammunition in a local controversy over spraying for mosquitoes, Carson promptly furnished him with a letter. She made a powerful case against spraying as self-defeating on two accounts: it killed off fish, birds, and other wildlife, contaminated vegetables and fruit, damaged flowers and shrubs, and poisoned soil and water; in addition, mosquitoes quickly developed resistance so that heavier spraying with ever more toxic substances would be necessary. "The irony of the situation," she concluded, "is that the harder we spray, the more rapidly we bring on the day when nothing we can use will reduce the mosquito population." Having heard Carson's arguments, the town board voted 4 to 1 against the spraying motion.[33]

When Carson returned to Maryland in September, a finished manuscript seemed in sight. She worked late into the night, focusing on what was her most difficult chapter, the discussion of synthetic chemicals. Somehow she had to strike a balance between the technical information necessary to explain how these chemicals worked and the need to simplify the material so that her readers could understand it. Of course, simplification increased the likelihood of introducing errors. But Carson excelled at this kind of writing—it was her gift, though one she employed only through painstaking effort. She also decided to open the

In the midst of her struggles with illness and the pressure to finish Silent Spring, *Carson found some peace at her home in Maine in the summer of 1961.* (Source: *Photograph by Bob Hines, courtesy of the Rachel Carson Council.*)

book with a fable rather than with a technical or political chapter, which might be too daunting to attract general readers. This literary strategy doubtless increased the appeal of the book.[34]

With her project on track, Carson suffered what she considered "the most frustrating and maddening" of all her ailments. In late November, she contracted a severe case of iritis, an inflammation of the irises in her eyes. For two weeks she was nearly blind. Reading was out of the question as even small amounts of light inflicted excruciating pain. Gradually, her sight returned; but over the next few months, her trusty assistant Jeanne Davis read the manuscript out loud so that Carson could make corrections. This was not the way in which Carson normally did revisions, and by January the whole affair was eating away at her: "If one were superstitious it would be easy to believe in some malevolent influence at work, determined by some means to keep the book from being finished." As a painstaking perfectionist, she wanted "to see [the manuscript], and on revision I have to keep going over and over a page—with my eyes!" That meant more struggle before she would let the book go to the printer. Despite all the obstacles, by late January she sent off the bulk of the manuscript to Rodell, Brooks, and Shawn. She likened her experience with the book to a dream in which "one tries to run and can't or to drive a car and it won't go." And yet, with the end in sight, she began to think about books she still hoped to write.[35]

For four years Carson had struggled with this project as it grew more cumbersome and technical. So even as she neared the end, doubts continued to plague her. How could someone who loved the beauty of nature convey that sense in a book about poisons? Who would want to read it? What words could make the story comprehensible, much less compelling? It was then that Shawn again came to her rescue. Carson trusted no one's literary judgment more than Shawn's. She could not mask her glee when she "*shamelessly*" repeated to Dorothy a few of his comments: "a brilliant achievement," "you have made it literature," "full of beauty and loveliness and depth of feeling." Equally reaffirming, Shawn wanted to publish that spring. She experienced an "enormous surge of relief—as if I now knew the book would accomplish what *I longed for it to do*."[36]

In writing *Silent Spring*, Carson built a formidable cadre of allies willing, even eager, to join her in battle. Her publishers at *The New Yorker* and Houghton Mifflin believed in her and in the importance of her message. Rodell worked closely with Carson's old friend, the Washington publicist Charles Alldredge, to go beyond the usual promotional sources. They sent copies of the book to opinion makers in Congress, government agencies, garden clubs, women's political organizations, and conservation groups such as the National Audubon Society. Even before excerpts appeared in *The New Yorker* in June, *Silent Spring* had attracted a wide and influential readership.

The completion of the book coincided with meetings of the Committee on Pest Control and Wildlife Relationships of the National Academy of Sciences–National Research Council. The council was a private, nonprofit research organization chartered to advise the government on scientific matters. Clarence Cottam, Carson's old friend from Fish and Wildlife days, had been promoting an investigation into the impacts of chemical pesticides on wildlife.

The committee published a report with three parts. The first two parts, "Evaluation of Pesticide–Wildlife Problems" and "Policy and Procedures for Pest Control," supposedly reflected a consensus of the scientific panel. In fact, several disgruntled members confirmed Cottam's suspicion that the committee's chair, W. H. Larrimer, revised the conclusions to favor his own views. Larrimer, Cottam noted, "spent most of his life in the USDA's Insect Control Division, and had worked especially on DDT and other chlorinated hydrocarbons. [Larrimer] was convinced beyond any question that no ill effects could result from the use of any of these chemicals." Still, even Cottam was shocked that parts one and two of the report supported the benefits of chemical pesticides without a hint of criticism.[37] He signed them nonetheless, to promote compromise within the committee. The third part Cottam refused to sign and thereby made public his dissent.

Cottam next turned to *Silent Spring* as another way to alert the public to what he, like Carson, saw as a looming environmental crisis. Carson had asked many scientific experts to check chapters for errors and misstatements. Most made few corrections and gave her unqualified endorsements but not Cottam. He wrote ten pages of comments about four wildlife chapters. He did so not because he objected to the material but because he was "convinced that you are going to be subjected to ridicule and condemnation by a few." Cottam's experience with Larrimer confirmed his belief that "facts will not stand in the way of some confirmed pest control workers and those who are receiving subsidies from pesticides manufacturers." He suggested places where Carson might qualify her views or offer more convincing evidence so that *Silent Spring* could make the strongest case possible against pesticides.[38]

Clarence Tarzwell, an aquatic biologist for the Public Health Service, served with Cottam on the Committee on Pest Control and Wildlife Relationships and was just as offended by the bias in its report. Like Cottam, Tarzwell believed *Silent Spring* might correct some of the misinformation the report had injected into the public debate. While he found nothing to correct in the chapter he reviewed for Carson, he did provide her with additional evidence that DDT accumulated in fish far off the Pacific coast.[39]

Ecologist Frank Egler spent much of his professional life struggling against an industrial–government alliance that promoted the use of herbicides along roads and highways. He too became a powerful Carson ally and gave her chapter "Earth's Green Mantle" an exquisitely thorough reading. In twelve dense pages of notes, he clarified her language and filled in omissions. Carson confessed that she was "overwhelmed at the thought of the hours you must have spent reading and commenting on my manuscript." Egler admired what Carson achieved, but even more he wanted to protect her from the torrents of abuse he was sure would follow.[40]

Scientists were not the only allies Carson made in preparing her book. Women in public life also championed her cause and none more ardently than Agnes Meyers, owner of the *Washington Post*. Meyers took an early interest in the book when she read an advance copy. In May, she hosted a luncheon to which she invited Washington's prominent female opinion makers, including a congresswoman, a senator, Frances Perkins, and the presidents of major women's organizations, including the League of Women Voters, the National Council of Jewish Women, and the American Association of University Women. These women, who shared Carson's passion for defending wildlife and nature, supported her as she came under fire from what turned out to be mostly male critics.

Conservationists offered Carson another body of allies. Supreme Court Justice William O. Douglas, who had himself written on pesticides, gave Carson more than one source she found useful. She, in turn, promised him an advance copy of the book. Douglas was one of a distinguished group of delegates who gathered in May 1962 for the White House Conference on Conservation, organized by Secretary of the Interior Stewart Udall, an early environmental advocate. At the conference, Carson met Udall as well as other prominent conservationists, some of whom had already read advance copies of her book. Udall assigned a member of his staff to track the book's reception and report ideas for future policy initiatives.[41]

Other signs indicated the book had wide appeal. The Book-of-the-Month Club made *Silent Spring* its October selection, and CBS approached Houghton Mifflin about doing a segment on the book for its highly rated news show *CBS Reports* (a precursor to *60 Minutes*). The National Audubon Society asked for permission to excerpt the book in a two-part series for its widely circulated *Audubon* magazine. Carson recognized that the book had "an irresistible initial momentum," and for a brief moment she declared herself "deeply and quietly happy."[42]

That moment of happiness did not last long. Shortly after the second installment of *Silent Spring* appeared in *The New Yorker*, the magazine's legal counsel received a phone call from the general counsel of Velsicol Chemical Company in Chicago, Louis McLean. McLean warned that Velsicol, as the sole manufacturer of heptachlor and chlordane, would sue if the magazine printed the last installment. Rodell and Carson worried about the possibility of lawsuits. Their worries only multiplied when Brooks determined that, while Houghton Mifflin was insured against libel, its authors were not similarly protected. Rodell quickly renegotiated Carson's contract to limit her exposure. Confident in this instance of the accuracy of Carson's material, *The New Yorker*'s counsel invited Velsicol to "go ahead and sue."[43]

Neither Brooks nor Carson learned of that threat until August, when Houghton Mifflin received a letter from McLean, this one implying that Velsicol planned a lawsuit if *Silent Spring* were published. Beyond this legal threat, McLean launched an attack that evoked the stridency of the anti-Communist senator Joe McCarthy. Chemicals were essential, he insisted, "if we are to continue to enjoy the most abundant and purest foods ever enjoyed by any country of this world." Beyond the "sincere opinions by natural food faddists, Audubon groups, and others," Mclean alluded to "sinister influences" attacking the chemical industry in the United States and Western Europe. Their ambition, he asserted, was first "to create the impression that all businesses are grasping and immoral" and, second, to reduce the use of agricultural chemicals "so that our supply of food will be reduced to east-curtain parity." "Sinister parties" inspired or financed the attacks even from the well-intended groups, McLean concluded ominously.[44]

What was it about *Silent Spring* and its author that produced such a furor? Carson did not simply indict DDT and dramatize the threat of pesticides to human health, though that was what she initially set out to do. As her moral outrage mounted, she quite self-consciously decided to write a book calling into question the paradigm of scientific progress that defined postwar American culture. She faulted the legion of scientists and corporate interests who, through arrogance or carelessness or willful ignorance, employed chemicals, a weapon "as crude as a cave man's club . . . against the fabric of life." "The current vogue for poisons," she argued, ignored the fact that modern technology was "dealing with life—with living populations and all their pressures and counter-pressures, their surges and recessions." The response from McLean and other spokespersons for the chemical industry suggested that they saw Carson for what she truly was— a subversive determined to crack the consensus from which the industry profited.[45]

Life against death was precisely Carson's point in opening the book with a grim tale that appealed to emotion. She created a dark fable of a mythical town in the heartland of America. Once upon a time, its people lived in harmony with their surroundings. The countryside teemed with wildlife and bountiful farms until a blight fell upon the land as if it were cast under an evil spell. Animals sickened and died, plants shriveled, and "everywhere was a shadow of death." People wondered where the birds had gone. Once lush roadsides "were now lined with browned and withered vegetation as though swept by fire." The only evidence to explain this disaster was a white powder that had fallen like snow several weeks before. This was "no witchcraft, no enemy action" that "had silenced the rebirth of life in this stricken world. The people had done it themselves." Carson assured her readers that such a town did not yet exist but suggested that it could someday have "a thousand counterparts in America or elsewhere in the world." She invited readers to wonder what "has already silenced the voices of spring in countless towns in America." She offered her book as an attempt to explain this mystery.[46]

What followed was a magisterial synthesis of a vast and disparate literature from government, scientific, and public health investigators. Carson built her case systematically, step by careful step. She opened the substantive chapters of the book with an idea she first introduced in *Under the Sea-Wind*: "The history of life on earth has been a history of interaction between living things and their surroundings." Where the environment molded life on earth, that life had only a minimal impact on the environment, or so Carson initially believed. But now she saw how a single species, humans, "acquired significant power to alter the nature of his world." The instruments of that power included "sinister" chemicals and radiation, which were irreversibly polluting air, earth, rivers, and seas. In that way, humans were mindlessly altering "the very nature of [the earth's] life." To drive home her point, Carson quoted Nobel Prize–winning humanitarian and naturalist Albert Schweitzer: "Man can hardly even recognize the devils of his own creation."[47]

Life as it existed took hundreds of millions of years to evolve until it "reached a state of adjustment and balance with its surroundings." Time was the crucial factor for future survival. "Given time," Carson wrote, "time not in years but in millennia—life adjusts, and a balance has been reached." Under the current chemical and nuclear assault, however, disruptive forces strike so rapidly that "in the modern world there is no time." The "impetuous and heedless pace of man" has overwhelmed "the deliberate pace of nature." Carson argued that no one could believe it was possible to "lay down such a barrage of poisons on the earth's surface without making it unfit for life." For that reason, "biocides" was a more accurate term than "insecticides." Worse yet, spraying was seldom the solution. Insects possessed a Darwinian capacity for survival, by which they sometimes underwent "flareback," returning in even greater numbers than before the spraying. "The chemical war is never won," Carson concluded, "and all life is caught in its violent crossfire." Why then incur this risk to life when the true problem in postwar America was not raising production but reducing surpluses the government stored at great public expense?[48]

Carson understood that in painting a bleak picture she risked alienating the very public she hoped to persuade. Thus, she stressed the need not to eliminate chemical pesticides but to understand the true problems posed by insects and the alternatives to indiscriminate spraying. Only with such an understanding could humans devise methods that did "not destroy us along with the insects." One problem arose from the extent of single-crop agriculture, a major source of insect explosions. Where nature maintained checks and balances through diversity, humans introduced vast tracts of uniform species. Dutch elm disease, for example, spread because so many cities lined their streets with a single variety of tree.

A second problem came from invasive species. Wittingly and unwittingly, humans introduced plant and insect pests with no natural enemies in the American landscape. Ecologist Charles Elton pointed the way to a nonchemical, biological means of managing this problem. Rather than target individual pests for eradication, humans needed to appreciate the relationship of living things to their surroundings in order to "promote an even balance and damp down the explosive power of outbreaks and new invasions." Like Elton, Carson saw biological controls as far superior to chemical ones because they employed nature's own defenses.[49]

Many scientific specialists and government officials thought in narrow, not holistic, ways. For them, each infestation demanded an immediate response; and since cost was a factor, that meant chemicals. Some zealots even envisioned a "chemically sterile, insect-free world." They launched their chemical assault without any oversight from enlightened authorities. Still, for Carson, the solution did not require a ban on pesticides, only their judicious use. She wanted them taken out of the hands of "persons largely or wholly ignorant of their potential for harm" and placed under the control of those experts who would weigh all the costs, environmental as well as economic. More important, Carson believed enormous numbers of people had been exposed to a chemical threat with neither their consent nor their knowledge. The public had both a right to know about the dangers pesticides posed and a role in the decisions to employ them.[50]

In a set of arguments anticipating later criticisms from government and industry, Carson charged that popular ignorance was no accident. Specialists who failed to educate the public were one source of the problem. The chemical companies were another, even more important, one. Determined to maximize profits, they fed the public "little tranquilizing pills of half truth," whenever evidence surfaced of harmful chemical applications.

Then, there were the insect controllers. In a world every day more reliant on technology, an uninformed public had to live with whatever these technocrats and bureaucrats calculated as beneficial. Yet those very authorities had a duty to inform the public before they acted. Only if people possessed the essential facts could they make an enlightened decision, and it was with ordinary people, Carson implied, that ultimate authority lay. Anticipating later movements calling for "power to the people," Carson quoted French philosopher Jean Rostand: "The obligation to endure gives us the right to know."[51]

Over most of its pages, *Silent Spring* provided what Carson saw as the essential facts. In her most technical chapter, "Elixirs of Death," she described the chemical structure of pesticides and explained how these poisons worked. Water, Carson informed her readers, was essential for more than human consumption, irrigation, and transportation. It played a vital role "in terms of the chains of life it supports." What happened, she asked, when poisons circulated in these cycles of nature? Like stealthy invaders, chemical poisons entered the lakes, streams, underground reservoirs, soil, and the plant life that covers the earth.

Carson then described all manner of events where pesticide applications went awry. In the case of water, for example, the story of Clear Lake in California was typically instructive. Clear Lake, 90 miles north of San Francisco, was a popular fishing and weekend vacation spot. It was also a breeding ground for the spectacular western, or "swan," grebe, whose nests floated on the lake's surface. Conditions on the lake also made it an ideal habitat for *Chaoborus astictopus*, a small, nonbiting gnat, annoying because of its dense concentrations. In 1949, using an approach that exemplified the technological prudence of the day, state officials decided to eliminate the gnat by spraying the lake with dichlorodiphenyldichloroethane (DDD), similar to DDT but less damaging to fish. Before application, the agency measured the volume of the lake to keep a ratio of one part DDD to every 70 million parts water. When the gnats revived in 1954, officials sprayed again with a slightly more concentrated ratio, after which they believed they had eradicated the problem.

Despite their caution, there were signs of trouble as hundreds of majestic grebes died. And by 1957 the gnats were back, followed by a third spraying and even more grebe deaths. When researchers thought to analyze fatty tissue in the dead birds, they discovered an amazing concentration of 1600 parts of DDD per million, even though spraying had never been at more than a 50 million-to-1 ratio. How had this happened? Examining fish offered a clue. Fish fed on plankton containing small amounts of DDD, and the grebe fed on fish. Lethal doses of the pesticide concentrated in their fatty tissues, even though the lake showed no residual trace of DDD. The DDD passed up the food chain, from the simplest organisms to the most complex, in growing concentrations.[52]

What did this mean for the Clear Lake fishermen who took home their catch for dinner? Carson warned her readers that experts linked these pesticides to increased risks of cancer. Despite its claim that DDD posed no hazard, the California Department of Public Health finally banned further applications in 1959. For Carson, this episode pointed to an essential ecological truth—"in nature nothing exists alone." All living things are connected through the web of life.[53]

The evidence mounted in succeeding chapters as Carson catalogued the killing of fish, the silencing of birds, the death of pets and farm animals, and even the accidental poisoning of workers who handled these deadly chemicals. In two of her most controversial chapters, she cited evidence that chemical carcinogens affected mitosis, or changes in the nucleus during cell division, the basic process of life. No laws or regulations required chemical companies to test their products to see if they altered chromosomes and, hence, the genetic inheritance of the entire human race.

In her last chapter Carson offered her readers a choice between the "smooth superhighway on which we progress with great speed" toward self-destruction and the less traveled road "that assures our preservation of the earth." The alternatives to the chemical road to perdition shared a common trait—they were biological approaches based on an understanding of the ecological niche in which the target organism existed. Most living things have natural enemies. Sterilization of the males of a species proved effective when the USDA employed it to control the screwworm plaguing cattle in Florida. Experiments with insect venoms, repellants, and hormones showed similar promise, as did bacteria, spores, fungi, and insects that attacked other insects. All that stood in the way of these new procedures, Carson asserted, was bureaucratic inertia and the determination of chemical companies to preserve their profits.[54]

Carson ended the book with her most telling blow against the culture of chemical controls. Using the metaphor of the cave dweller, she described the modern notion of "the control of nature" as "a phrase conceived in arrogance, born in the Neanderthal Age of biology and philosophy, when it was supposed that nature exists for the convenience of man." Then Carson identified her main villains in this morality tale—applied entomologists, whose concepts and practices, she charged, "date from the Stone Age of science." So narrow was their approach to insect control that they favored economic considerations over environmental ones. To drive home her point, she concluded with an ominous observation: "It is our alarming misfortune that so primitive a science has armed itself with the most modern and terrible weapons, and that in turning them against the insects it has also turned them against the earth."[55]

Carson's critics attacked both her message and the messenger. Not surprisingly, entomologists were among her most vocal assailants. Economic entomologists felt themselves under direct assault as the narrow-minded experts whom Carson charged with seeking only to improve the productivity of the agricultural economy. Indeed, in the 1920s, they did determine that chemicals offered both an effective and an inexpensive means to control insect pests. In the 1930s, DDT became their atomic bomb, though they believed that, unlike its nuclear cousin, it posed no threat to humans and a minimal danger to wildlife. In condemning DDT and the science that promoted its use, Carson was impugning their personal and professional reputations. Many entomologists dismissed *Silent Spring* as an emotional rant, hardly the factual indictment of a responsible scientist. George Decker, an economic entomologist and former USDA advisor, charged that "*Silent Spring* poses leading questions, on which neither the author nor the average reader is qualified to make decisions. I regard it as science fiction, to be read in the same way that the TV show *Twilight Zone* is to be watched." More broadly, the magazine *Science* concluded "Rachel Carson's stretching of scientific points is not easily excused. . . ."[56]

Where the entomologists were personally and professionally offended, the chemical companies and agribusinesses feared major financial losses if the public heeded Carson's warning. As *Aerosol Age*, a trade journal, warned, "For the insecticide industry, this book could turn out to be a serious and costly body blow—even though it did land below the belt." The reviewer concluded, "All the case histories fall on the debit side."

In similar ways, other industry critics failed to confront Carson's argument that pesticides damaged wildlife and posed a significant risk to public health. Instead, they launched a publicity campaign in which they invited the public to imagine a world without chemicals—a position Carson never advocated. Carson, they suggested, wanted to return the world to a state of hunger and want. If Americans accepted her ideas, warned William Darby, head of biochemistry at Vanderbilt School of Medicine, they would face "the end of all human progress, reversion to a passive social state devoid of technology, scientific medicine, agriculture, sanitation. It means disease, epidemics, starvation, misery, and suffering." Darby, who had served on the Committee on Pest Control and Wildlife Relationships of the National

Academy of Sciences–National Research Council and supported the reports that so enraged Clarence Cottam, was one writer who dismissed Carson as simply hysterical.[57]

A third group of Carson's critics were the apostles of technology, who saw her as an enemy of scientific progress and the "American way of life." Officials from the Public Health Service maintained that their research demonstrated that DDT and other pesticides did no lasting damage to birds, fish, or other wildlife and was harmless so long as trained persons exercising reasonable caution applied them. Indeed, officials claimed that virtually no evidence existed of any case of DDT poisoning. An impressive array of scientists supported the industry's case for the judicious use of chemicals. Those who thought otherwise, argued Darby (who identified himself with this group as well), were "the organic gardeners, the anti-fluoride leaguers, the worshippers of 'natural foods,' those who cling to the philosophy of a vital principle, and other pseudo-scientists and faddists." He and Louis McLean of Velsicol were not the only critics who saw Carson as an agent of some subversive cult. Ezra Taft Benson, a Mormon elder who served as President Eisenhower's secretary of agriculture, turned viciously *ad hominem* when he wondered "why a spinster with no children was so concerned about genetics?" He added his voice to those who saw Carson as an agent of subversion by concluding that she was "probably a Communist."[58]

The notion that rational scientists supported chemical pesticides and only subversives, cranks, and faddists opposed them anticipated a fault line that would widen in the 1960s. That line divided the worldviews of empirical and holistic scientists, of the new ecologists and traditional conservationists, and of the champions of consensus and conformity and countercultural advocates of life in harmony with nature. The tensions between those points of view played into the controversy over *Silent Spring*. More than the use and abuse of pesticides was at stake. Most traditionalists believed that by attacking *Silent Spring* they were preserving the basis of the American way of life. Those who would create the new environmental movement were similarly convinced that in defending *Silent Spring* they were promoting a better world.

The groups seeking to discredit Carson and her book were generally well organized and financed. Chemical company publicists, for example, could enlist a wide range of scientists and public figures to make their case. By contrast, Houghton Mifflin had few resources and more limited access to the media with which to counteract the hostile commentary. Nor could Carson and her supporters respond to every attack against her. Yet, ironically for the industry, the more chemical company spokesmen raged against Carson, the more publicity they generated for *Silent Spring*. By December, barely two months after its publication, bookstores had sold over 100,000 copies. The book stood at number one on the bestseller list, and even more important, the controversy had become news. Determined to discredit Carson, her adversaries helped spread her message and in the process made her a celebrity. At an August news conference, a reporter asked President Kennedy if the USDA and Public Health Service were investigating the danger of pesticides. Kennedy assured him that they were and added "particularly since the publication of Miss Carson's book."[59]

Time and *Life* magazines, publications of Henry Luce's Time-Life publishing empire, generally supported big business and the Republican Party; but they split their decision on Carson and her book. On September 28, one day after the official publication of *Silent Spring*, the science correspondent for *Time* damned Carson with the faintest praise by suggesting she had put "her literary skill second to the task of frightening and arousing her readers." The book was rife with errors, simplifications, and exaggerations, he charged. With condescending charity, he allowed that many scientists shared her "love of wildlife, and even her mystical attachment to the balance of nature" but concluded that those same scientists "fear that her emotional and inaccurate outburst in *Silent Spring* may do harm by alarming the non-technical public, while doing no good for the things she loves."[60]

Life magazine proved far more sympathetic. To illustrate its article, "The Gentle Storm Center: A Calm Appraisal of *Silent Spring*," it assigned the twentieth century's leading photojournalist, Alfred Eisenstaedt, to shoot the photographs. Jane Howard, *Life*'s interviewer, described Carson as a "formidable adversary" but not a fanatic. Carson assured Howard she was on no "Carrie Nation Crusade," a reference to the axe-wielding temperance reformer. She simply feared "the next generation will have no chance to know nature as we do—if we don't preserve it the damage will be irreversible." Inevitably, Howard commented on the controversy between Carson and the chemical interest but, on the whole, commended Carson's point of view. Howard also addressed an issue Carson's adversaries constantly raised—her gender. Carson explained that, while she was unmarried, she was not a feminist. She was interested not in "things done by women or men, but in things done by people."[61]

Critics only increased their volume as the book gained more attention. Most formidable among them was Dr. Robert White-Stevens, the assistant director of the Agricultural Research Division of American Cyanamid, a major chemical producer. Tall, imposing in manner, English by background, and possessed of a sly sense of humor, he described Carson as "a fanatic defender of the cult of the balance of nature." That charge was common among empirical scientists, who routinely dismissed holistic ecologists such as Carson as more mystic than scientific. While Carson questioned the use of pesticides, White-Stevens believed they were essential and safe. In perhaps his most telling point, he argued that restrictions on pesticides would increase hunger and disease in the developing world. "Chemicals offer the only immediate hope of increasing food production to meet world needs," he steadfastly maintained.[62]

Carson more than held her own against the onslaught, but the need to respond to endless misinformation, vituperation, and distortion wore on her. Some of her most outspoken critics had not even read her book or chose to ignore her qualified condemnation of chemical pesticides. In a speech to the Women's National Press Club in December 1962, Carson described the response of officials from two Pennsylvania county farm bureaus, both outraged by and largely ignorant of the content of her book: "No one in either county farm office who talked to us had read the book, but all disapproved of it heartily." While she retained her sense of humor, Carson did confess "The struggle against the massed might of industry is too big for one or two individuals . . . to handle."[63]

The battle she was waging was bigger than almost anyone knew. Her cancer was spreading almost as rapidly as the controversy. During the final preparation of the manuscript in the spring of 1962, her doctors discovered that her cancer had metastasized to the right side of her body. The "two million volt monster" X-ray machine was her "only ally . . . but what an awesome and terrible ally, for even while it is killing the cancer I know what it is doing to me." Carson bought a wig to mask the most obvious evidence of her radiation treatments, but it did nothing to relieve the nausea, vomiting, and fatigue that inevitably resulted.[64]

As her hair fell out and her energy waned, the attacks on her book and on her character flowed in a torrent. So did requests for interviews and speeches and, to her utter delight, the list of awards her book was winning. She refused most interviews but made an exception for *CBS Reports*, one of the most popular interview shows on television. Carson saw her appearance as an opportunity to tell her story to millions of Americans during primetime. Since she was too ill to go to the studio, the show and its host, Eric Sevareid, came to her. As the television cameras opened her home to the nation, a pale and demure Rachel Carson entered the homes of Americans. She answered Sevareid's questions with calm self-assurance, confident that after all the years of research the facts would make her case. Neither Sevareid nor Jay McMullen, the show's producer, had any doubt that Carson was the master of her material—or that she would live very much longer.[65]

In December, jostled by a crowd of Christmas shoppers as she looked for Roger's present, Carson fainted for the first time in her life. Her doctor blamed stress, but her oncologist knew better. Many patients undergoing radiation treatments develop angina, a sharp, vice-like pain in the chest, brought on by the weakening of the heart muscles. The damaged heart struggles to deliver oxygen to the brain. Even if cancer did not kill her, a heart attack certainly could. That episode brought 1962 to a somber end. In her Christmas letter to Dorothy, Rachel wrote of her "joy and fulfillment . . . in *Silent Spring* being published and making its mark," on the one hand, and of "the shadow of ill health," on the other. That shadow was lengthening and cast a pall over the future, but it did not diminish her determination to fight on.[66]

Yet, all her righteous fervor could not guarantee that the public would respond to her warning. Her message was not a popular one, and the forces aligned against her included leaders of science, industry, and government. Despite the power of her adversaries, Carson managed to win the battle, though it remained unclear who won the war. On the road to vindication, Carson won an early victory over her nemesis, Dr. Robert White-Stevens, the spokesperson for the chemical industry. *CBS Reports* again provided the arena, this time for a showdown on a program entitled "The Silent Spring of Rachel Carson." The show, eight months in the making, included excerpts from the interview at Carson's home the previous November as well as appearances by White-Stevens and representatives from the government.

As the April 3 broadcast date neared, Carson became increasingly anxious, especially when she read over the list of people scheduled to appear with her on the program. She couldn't help feeling that the show was stacked with her detractors. Nor was she confident that she was physically up to the challenge. In the spring of 1963, she was in the middle of radiation treatments and felt totally exhausted. Privately, she hoped she didn't "look and sound like an utter idiot" when the cameras began to broadcast. To add to the tension surrounding the show, three sponsors, a home products company and two food producers, pulled their advertising before airtime because they thought the content was too controversial.[67]

On the day before the broadcast, Carson received an unexpected boost from *New York Times* critic Brooks Atkinson, already a fan of her work. In August 1962, Atkinson had defended her as "a biologist and writer" who "understands that chemical sprays are a permanent part of technology. But she hopes people will learn to use them intelligently with a knowledge of the deadly peripheral damage they can do to the balance of nature." Having made that point, Atkinson was offended when critics from the chemical industry, White-Stevens among them, denounced her as "a fanatic defender of the cult of the balance of nature." To that, he scoffed in his April 2, 1963, column "as if the balance of nature were equivalent to nudism or skin diving." Atkinson concluded with an unequivocal endorsement, "The evidence continues to accumulate that she is right and that *Silent Spring* is the 'rights of man' of this generation."[68]

As the April 2 installment of *CBS Reports* flickered onto TV screens, producer Jay McMullen gave Carson ample time to explain her views and White-Stevens an equal opportunity to rebut them. Eric Sevareid provided the opening commentary. In his calm and authoritative voice, Sevareid assured viewers that a pesticide problem did exist and emphasized that Carson favored other means of control and only a gradual phasing out of chemical pesticides. He turned to Carson. Rachel was about to have her day in the court of public opinion.

Anyone expecting a fanatical mystic was surely disappointed, for there on the screen sat an attractive woman in a dark dress set off with a gold necklace. Those aware of her condition noticed, too, the puffiness in her face—a result of radiation treatments. Carson made her case with reasoned arguments delivered in a dignified manner. Having heard only about the benefits and safety of pesticides, she explained, the public knew "very little about the hazards, very little about the failures." Her goal was not to alarm but to ensure that the public "being asked to acquiesce in their use" had the facts needed to make informed decisions. She then read excerpts from *Silent Spring* that documented the hazards of pesticide use.[69]

Next came a parade of government officials interviewed by producer McMullen. They virtually made Carson's case for her. More befuddled than enlightened or enlightening, few of them expressed any sense of the potential hazards chemicals posed or of any plans to subject them to closer regulation or testing. Worse yet, they seemed unconcerned about the threat to human health or the environment. But if anyone truly frightened the public, it was Dr. White-Stevens. On television his assured tone came across as arrogant, his suave manner as oily. In ominous tones he described the plagues awaiting a world without chemical pesticides.

"If man were to faithfully follow the teachings of Miss Carson," he warned the audience, "we would return to the Dark Ages and the insects and diseases and vermin would once again inherit the earth."[70]

The final words belonged to Carson, who used them to express her ecological vision of the relationship of humans and nature. "We still talk in terms of conquest," she observed. "We still haven't become mature enough to think of ourselves as only a tiny part of a vast and incredible universe." Without hesitating, she delivered her final blow: "I think we're challenged as mankind has never been challenged before, to prove our maturity and our mastery, not of nature, but of ourselves." The challenge echoed John F. Kennedy's stirring call to new frontiers.[71]

Carson's friends and allies applauded her performance, convinced she had persuaded some 10 to 15 million viewers that her work was sound. "How delightful that Dr. White-Stevens looked so fiendish," one friend exalted, while biologist Frank Egler said, "You scored a notable triumph!" The following day, the true nature of her triumph became apparent. Addressing his Senate colleagues, Abraham Ribicoff of Connecticut explained that "last night's CBS telecast clearly showed there is an appalling lack of information on the entire field of environmental hazards." Ribicoff announced that his Government Operations Subcommittee would hold hearings. The standout among the witnesses called to testify would be Rachel Carson.[72]

An even more affirming moment awaited. On May 15, the day Ribicoff's subcommittee held its first hearing, the President's Science Advisory Committee (PSAC) issued its report "Use of Pesticides." Having already testified before the PSAC, Carson feared that friends of the chemical industry on the committee would dilute the report and its impact, just as they had commandeered the National Academy of Sciences–National Research Council report on pest control a year earlier. Rumors at first deepened her fears, but on the evening before publication Carson learned the PSAC report stated unequivocally "the accretion of residues in the environment can be controlled only by the orderly reductions of persistent pesticides." In what amounted to a personal coup for Carson, the report noted "until the publication of *Silent Spring*, people were generally unaware of the toxicity of pesticides."[73]

To Carson's surprise, the PSAC believed that during the hearings USDA officials were obstinate and chemical industry witnesses evasive. Agriculture officials drew a line in the sand. For them it was either all-out use of pesticides or utter disaster. They simply refused to discuss a more restrictive approach to the use of chemicals. Industry witnesses hid behind regulations. They claimed to have met USDA and FDA safety standards and worried that stiffer regulations might discourage the development of new products. Neither government nor industry witnesses would admit, however, that pesticides posed any hazard.

The committee members rejected those arguments. They recommended more extensive monitoring of pesticide residues, increased research into chemical pesticide toxicity, development of alternative methods of pest control, and more open documenting of control programs undertaken. Above all, they concluded, "The elimination of the use of persistent toxic pesticides should be the goal." Carson told a correspondent from CBS "that the report has vindicated me and my principal contentions."[74]

Rachel Carson (far left) meeting with the President's Science Advisory Committee in the spring of 1963. Carson is conspicuously the only woman present. (SOURCE: Courtesy of the Lear/Carson Collection, Connecticut College.)

The news media agreed with her assessment. "RACHEL CAR-SON STANDS VINDICATED," headlined the May 15 *Christian Science Monitor.* Eric Sevareid told his *CBS News* audience that Carson had set off with two goals—"to alert the public" and "to build a fire under the government." Through *Silent Spring* she achieved her first goal. "Tonight's report by the presidential panel is *prima facie* evidence that she has also accomplished the second." Both *Science* magazine and *Chemical and Engineering News,* having published hostile reviews of *Silent Spring,* now admitted that the report supported Carson and her thesis.[75]

Surely the most gratifying response came from Dorothy Freeman. She pointed out that five and a half years of labor "had reached a positive conclusion today" and added "I doubt you could have summed it up any better than did Eric Sevareid—'a voice of warning and a fire under the government.'" Dorothy appreciated that a woman had written *Silent Spring,* a title that to her sounded more feminine and thus less confrontational than *Man Against the Earth.* She predicted that Rachel's name would be remembered when the world had forgotten Gordon Cooper, the second Mercury astronaut circling the earth the very day she wrote her note. "And poor Gordon is still out in space," Dorothy joked.[76]

Humor of another kind helped sell the book abroad, where it was translated into twenty-two languages. Lord Shackleton, son of the famed Antarctic explorer and a dedicated conservationist, wrote the introduction to the English edition. With dry wit, Lord Shackleton rose grandly to his feet in the House of Lords to inform the members that Polynesian cannibals now refused to eat Americans "because their fat is contaminated with chlorinated hydrocarbons." Such bad fortune for the Americans afforded the English an excellent opportunity to expand their trade, for, as he jested, "we are rather more edible than Americans . . . that we have about 2 parts per million of DDT in our bodies, whereas the figure for Americans is about 11 million parts per million." Inspired, perhaps, in small degree by Shackleton's wit and more by *Silent Spring,* Britain soon tightened regulations on pesticides.[77]

Carson's reputation grew to near legendary heights. When she appeared before Ribicoff's Senate subcommittee in May 1963, the senator echoed Abraham Lincoln's oft-quoted remark to

Harriet Beecher Stowe, the author of *Uncle Tom's Cabin,* a book often credited with exposing the evils of slavery, "You are the lady who started all this." The tribute, even if a bit patronizing, was certainly warranted. Her writing did introduce Americans to the ideas of environmentalism and the need to live in harmony with nature. As she leaned toward a bank of microphones, she told the senators of new evidence, collected in the past two years, indicating that pesticide pollutants were spreading from areas of their application to remote regions of land and sea. She repeated her argument that each citizen had a right "to be secure in his own home against the intrusion of poisons applied by other persons," the same issue that had ignited the Long Island lawsuit five years earlier. Overwhelming evidence made clear that pesticides should be strictly controlled and eventually eliminated. After an hour and a half of questions in the packed hearing room, Carson was ready to leave but not before Senator Ribicoff commended the strength of her convictions and her iron-willed determination to promote the public welfare.[78]

Carson testified again several days later before the Senate Committee on Commerce, but she made few public appearances after that. By the summer of 1963 she knew she was dying. Each meeting with friends, each sunrise, or her annual trip to Maine, she realized, might be her last. More than anything, she wanted to escape to Southport and Dorothy. "There must be many opportunities for us to do together the things we love to do," she wrote her friend.

By June her health had deteriorated to such a point that it was not clear she could manage a summer at Southport with only young Roger as company. Marie Rodell persuaded Rachel's old friend Lois Crisler to stay with her. Rachel soothed Dorothy's disappointment at this complication by pointing out that without Lois "I wouldn't have gotten to Maine at all." More often now, angina sent flashes of pain through her chest. Compression fractures in her spine, another side effect of her radiation treatments, made walking both difficult and dangerous: "A fall, I have been warned, could have extremely serious results." No matter what the risk, Rachel was determined to return to the place and the person who meant the most to her.[79]

The summer in Southport with its tide pool creatures, shadowy woodlands, and ocean breezes passed all too quickly. Despite all her health complications, Carson managed to return to her writing, keep up her correspondence, and continue the crusade she began with *Silent Spring*. New honors poured in. Among them, two had special meaning. In January 1963, the Animal Welfare Institute awarded her its Albert Schweitzer medal. Having dedicated *Silent Spring* to the philosopher, humanitarian, and animal-rights activist, Carson could imagine no finer tribute. His notion of the "reverence for life" had been a cornerstone of her own environmental thinking. From Dr. Schweitzer she had learned "we are not being truly civilized if we concern ourselves with the relation of man to man. What is important is the relation of man to all life." Carson devoted her gifts as a writer to spreading that message.[80]

In December 1963, Carson won three additional honors. The National Audubon Society chose her as the first woman to receive its medal. Two days later the American Geographical Society gave her its Cullum Medal. At each ceremony, she spoke about the interdependence of humans and nature and the respect for life Albert Schweitzer had inspired. But the honor that moved her more than any other was her election to the American Academy of Arts and Letters. Limited to just 50 members, the academy included the nation's artistic and literary elite. Carson was one of only four women inducted into that august body and the only writer of nonfiction among them. She spoke of her election as "the most deeply satisfying thing that had happened in the honors department." Comparing her to Galileo and Buffon, academy president Lewis Mumford, one of the nation's leading critics and social thinkers, credited Carson with quickening "our consciousness of living nature and [alerting] us to the calamitous possibility that our shortsighted technological conquests might destroy the very source of our being." There in the audience to share in the pride and poignancy of the moment sat friends who had meant so much to her life and her writing, Marie Rodell and Stan and Dorothy Freeman.[81]

A month earlier, in November 1963, an event vindicated Carson in the starkest terms imaginable. Strollers along the Mississippi River levees discovered masses of dead fish floating on the surface. Louisiana officials were no strangers to such mysterious fish kills; they had witnessed some thirty over the past decade alone. But this one was different. Never before had they seen so many dead fish. Louisiana's Office of Water Pollution Control contacted the Public Health Service in Washington once they realized that natural causes could not explain the epidemic deaths. Investigators ruled out parasites and disease. After painstaking

analysis using state-of-the-art gas chromatography, they identified the culprit as almost undetectable amounts of the chemical
pesticide endrin.

What was the source of the endrin? Investigators finally linked
it to a waste-treatment plant in Memphis. In April 1964, Paul
Brooks informed Carson that the operator of the plant was none
other than Velsicol, the very company that had tried to block the
publication of *Silent Spring*. As much as anyone, Carson was responsible for this remarkable discovery. "How does Rachel
Carson look now?" a reporter asked public health officials. "She
looks pretty good," one of them answered. The incident led
Senator Ribicoff to open a new round of Senate hearings and,
partly as a result, introduce a new clean water bill.[82]

By the time she received the news about Velsicol, Carson was
too sick to enjoy the irony of the company's fall from grace.
Cancer had spread to her brain. Writing to Rodell over the spring
of 1964, she had faced up to that lurking threat: "I have in mind
that if all these recent tests showed any invasion of the brain, I
would just cancel my outstanding engagements for I would
rather not scatter my remaining time but use it exclusively for
writing." It was not to be; the voice that had awoken the nation
fell silent. Dorothy came in April for a final visit, though by then
Rachel only occasionally recognized that she was there.[83]

On April 14 Carson suffered a fatal heart attack. By then she had
reconciled herself to death and said her good-byes to Dorothy
and other friends. Fittingly, an ecological vision sustained her as

the end approached. In a letter to Dorothy written after her final visit to Maine, she recalled the flight of monarch butterflies setting off on a journey from which they would not return. Neither she nor Dorothy felt sadness at the sight since they understood "when any living thing has come to the end of its life cycle we accept that end as natural." To this thought Carson added her own notion of life and death:

> For the monarch, that cycle is measured in a known span of months. For ourselves the measure is something else, the span of which we cannot know. But the thought is the same: when the intangible cycle has run its course it is a natural and not unhappy thing that a life comes to its end.[84]

So it was for Rachel Carson. From her earliest days she was drawn, as if by some invisible force, to the study of nature. All forms of life, she came to believe, were connected in an endless cycle. In her writing, she dedicated herself to revealing this vision. When the hubris of humankind endangered nature with chemicals so toxic they threatened life itself, she sounded the alarm. Her dedication as a scientist made her credible; her gifts as a writer made her inspirational. Pilloried and honored, she and her work breathed life into a fledgling environmental movement determined to protect the living things she loved so dearly. Only time would tell whether it would succeed.

NOTES

1. Carson to Lois Crisler, February 8, 1962, RCP-BLYU.
2. Carson to Margaret Minghan, January 1, 1960, RCP-BLYU; see also Brooks, *House of Life*, p. 245.
3. Russell, *War and Nature*, pp. 165–203.

4. Ibid., pp. 185–187.

5. For an edifying and entertaining look at the CIA and LSD, see Martin Lee and Bruce Shlain, *Acid Dreams: The CIA, LSD, and the Sixties Rebellion* (New York, Grove Press, 1985).

6. Russell, *War and Nature*, p. 219; Graham, *Since Silent Spring*, p. 32; Lear, *Rachel Carson*, pp. 317–326.

7. John Kenneth Galbraith, *The Affluent Society* (New York, Mentor, 1958), p. 31.

8. Ibid., pp. 197–203.

9. Eisenhower's speech is widely available online; see, for example, www.eisenhower.archives.gov/farewell.htm.

10. Dunlap, *DDT*, pp. 104–105.

11. Lytle, *America's Uncivil Wars*, p. 112.

12. Lear, *Rachel Carson*, p. 360; RC to DF, November 19, 1959, in Freeman, *Always, Rachel*, pp. 290–291; and Dunlap, *DDT*, pp. 107–108.

13. Dunlap, *DDT*, pp. 104–105.

14. Lear, *Rachel Carson*, p. 347.

15. *Washington Post*, Dec. 2, 1958; Carson to Marjorie Spock, December 4, 1958, RCP-BLYU.

16. Lear, *Rachel Carson*, p. 119; Carson to Harold Lynch, July 15, 1945, RCP-BLYU.

17. Lear, *Rachel Carson*, pp. 342–343; Carson to Clarence Cottam, November 18, 1958, Spock to Carson, March 3, 1959, and Cottam to Carson, November 18, 1959, RCP-BLYU.

18. Lear, *Rachel Carson*, pp. 342–344.

19. Ibid., pp. 354–356.

20. Carson to Paul Brooks, December 3, 1959, RCP-BLYU.

21. RC to DF, December 29, 1959, in Freeman, *Always, Rachel*, p. 295.

22. Ibid., DF to RC, January 9, 1960, pp. 296–297.

23. Ibid., RC to DF, February 10, 1960, p. 302; Carson to Brooks, April 1, 1960, RCP-BLYU.

24. Carson to Spock, April 12, 1960, RCP-BLYU.

25. Lear, *Rachel Carson*, pp. 369–370.

26. Carson to Spock, May 18, 1960, RCP-BLYU.

27. RC to DF, in Freeman, *Always, Rachel*, September 8, 1960, p. 309, and October 12, 1960, pp. 310–311.

28. Lear, *Rachel Carson*, p. 376.
29. Marie Rodell to Carson, December 3, 1960, RCP-BLYU.
30. Brooks, *House of Life*, pp. 268–269.
31. Lear, *Rachel Carson*, p. 380; Carson to Brooks, December 27, 1960, RCP-BLYU.
32. RC to DF, in Freeman, *Always, Rachel*, February 2, 1961, pp. 337–338; February 24–25, pp. 351–352; March 2, 1961, p. 354; March 4, 1961, p. 354–355, and March 20, 1961, p. 363. See also Brooks, *House of Life*, p. 272.
33. Brooks, *House of Life*, pp. 373–374; Carson to Crisler, September 19, 1961, RCP-BLYU.
34. Lear, *Rachel Carson*, pp. 392–393.
35. RC to DF, January 6, 1962, in Freeman, *Always, Rachel*, pp. 390–391; see also Brooks, *House of Life*, pp. 276–277.
36. Brooks, *House of Life*, p. 277; RC to DF, January 23, 1961, in Freeman, *Always, Rachel*, pp. 393–394.
37. Graham, *Since* Silent Spring, pp. 48–50.
38. Cottam to Carson, February 14, 22, and 26, 1962, RCP-BLYU.
39. Clarence Tarzwell to Carson, August 31, 1962, RCP-BLYU.
40. Frank Egler to Carson, January 26, 1962, and Carson to Egler, January 29, 1962, RCP-BLYU.
41. Lear, *Rachel Carson*, pp. 405–407.
42. RC to DF, June 13, 1962, in Freeman, *Always, Rachel*, p. 407.
43. Lear, *Rachel Carson*, p. 418; Brooks, *House of Life*, pp. 298–299.
44. Graham, *Since* Silent Spring, p. 60.
45. Carson, *Silent Spring*, pp. 296–297.
46. Ibid., pp. 1–3.
47. Ibid., pp. 5–6.
48. Ibid., pp. 8–9.
49. Ibid., pp. 10–11, 117.
50. Ibid., p. 12.
51. Ibid., pp. 12–13.
52. Ibid., pp. 46–50.
53. Ibid., pp. 186–216.
54. Ibid., p. 279.
55. Ibid., p. 297.
56. Dunlap, *DDT*, p. 112; reviews collected in RCP-BLYU.

57. Dunlap, *DDT*, pp. 112–113.
58. *Chemical and Engineering News*, October 1, 1962; Dunlap, *DDT*, p.106.
59. Graham, *Since* Silent Spring, p. 61.
60. *Time*, September 28, 1962.
61. *Life*, October 12, 1962, pp. 105–106 and 109–110; RC to DF, October 4, 1962, in Freeman *Always, Rachel*, pp. 410–411.
62. White quoted in Easton, PA, *Express*, September 29, 1962, RCP-BLYU.
63. Carson, speech to the Women's National Press Club, December 5, 1962, and Carson to Cottam, January 1, 1963, RCP-BLYU.
64. RC to DF, in Freeman, *Always, Rachel*, March 28, 1962, p. 399, and April 10, 1962, p. 403.
65. Lear, *Rachel Carson*, p. 425.
66. RC to DF for Christmas, December 1962, in Freeman, *Always, Rachel*, p. 420.
67. RC to DF, in Freeman, *Always, Rachel*, March 26, 1963, p. 445, and April 1, 1963, p. 451; see also Graham, *Since* Silent Spring, pp. 82–83.
68. For Atkinson's columns, see *New York Times*, September 11, 1962, and April 2, 1963.
69. RC to DF, April 1, 1963, in Freeman, *Always, Rachel*, p. 451. Some footage from the program is available on *The American Experience*, "Rachel Carson's Silent Spring," WGBH–TV, Boston, WNET–TV, New York, and KCET–TV, Los Angeles, 1992, distributed by PBS Video. See also *CBS Reports*, "The Silent Spring of Rachel Carson," April 3, 1963, transcript.
70. *CBS Reports*, "Rachel Carson."
71. *CBS Reports*, "Rachel Carson."
72. Christine Stevens to Carson, April 11, 1963 and Egler to Carson, April 4, 1963, RCP-BLYU.
73. "Use of Pesticides: A Report of the PSAC" (Washington, D.C., GPO, 1963), p. 23.
74. Graham, *Since* Silent Spring, p. 84.
75. Ibid., p. 85; see also reviews in RCP-BLYU.
76. DF to RC, May 15, 1963, in Freeman, *Always, Rachel*, p. 461.
77. The Shackleton anecdote is told in Brooks, *House of Life*, p. 317.

78. Worster, *Nature's Economy*, p. 350; see also Lear, *Rachel Carson*, p. 454; Al Gore, introduction to *Silent Spring*, p. xix; Graham, *Since Silent Spring*, pp. 81–82 and 91–92.

79. RC to DF, June 1, 1963, in Freeman, *Always, Rachel*, p. 465.

80. Brooks, *House of Life*, p. 321.

81. Ibid., p. 327; Lear, *Rachel Carson*, fn. 50 p. 582.

82. Graham, *Since Silent Spring*, pp. 96–109; Brooks, *House of Life*, p. 329.

83. Carson to Rodell, spring 1964, RCP-BLYU.

84. RC to DF, September 9, 1963, in Freeman, *Always, Rachel*, p. 467.

Epilogue

Rachel Carson: The Legacy

THE FUNERAL WAS NOTHING LIKE THE ONE RACHEL HAD planned. She had requested a simple service at All Souls, a Unitarian church where Duncan Howlett served as minister. Howlett comforted Rachel in her last months and helped her plan a foundation that became the Rachel Carson Trust for the Living Environment. In keeping with her love for the seashore, she asked to have her body cremated and her ashes spread near her cottage in Maine.

When the time came for the service, however, Howlett was not involved. Rachel's brother Robert, who'd had little time for the family when Rachel was alive, instead planned a funeral full of pomp and ceremony. The service took place at the mammoth, partially completed Washington National Cathedral. Episcopal Bishop William Creighton, a clergyman Carson scarcely knew, presided as some 150 friends and notables gathered to pay her tribute. The choice of honorary pallbearers was the only fragment left of Rachel's original plan, but only because Marie Rodell arranged that part of the service. Since the pallbearers were a distinguished group—Senator Abraham Ribicoff, Interior Secretary

Stewart Udall, Robert Cushman Murphy, Edwin Way Teale, Charles Callison of the National Audubon Society, and Bob Hines, the illustrator for *The Edge of the Sea*—Robert did not object. Still, there were no memorials from those who knew her or her work. Robert thought it more fitting that Bishop Creighton offer a prayer for those who had died at sea.

In almost every way, Robert turned Rachel's death into an ordeal for those dear to her. He did not even bother to call her closest friends, most of whom learned from the radio or the morning papers that she had died. Nor did he agree to the cremation or the spreading of her ashes in Maine. Instead, he arranged with a local funeral home to have her casket open for the viewing of her body, later to be buried in the plot next to her mother. Friends sensitive to her wishes forced Robert into a compromise—half her ashes would be buried and Dorothy would scatter the remainder the following summer at Southport. To avoid any embarrassing revelations, Robert rifled through Rachel's personal papers and destroyed those he found sensitive. When he discovered Rodell gathering Rachel's manuscript materials, he ordered her out of the house.

Then there was Roger. Robert could barely conceal his dislike for the living evidence of his niece's sinfulness. The day after Rachel died he came upon the young boy watching television in his room at her house. The set, a gift from Rachel, offered Roger a small measure of comfort at this sad time. Informing the frightened young boy that all the furnishings in the house now belonged to him, Robert removed the television, leaving Roger with no idea what life had in store. Prior to her death Rachel remained uncertain of how to arrange the care and guidance Roger needed, though in her will she provided for him financially. In a codicil, she mentioned Dorothy's son, Stan Jr., and Paul Brooks as possible guardians. Stan Jr. reluctantly declined, but Brooks and his wife Susie agreed to take Roger into their family.

Rachel's friends found their own way to honor her last wishes. On Sunday April 17, two days after the extravagant ceremony at Washington National Cathedral, they gathered at All Souls for the Sunday service. No ornate stonework and woodcarvings decorated this modest church. The spirit of the day lay in the special connection the worshippers felt for their departed friend. Dorothy, Jeanne Davis, Marie Rodell, and Bob Hines paid their last respects. Rev. Howlett then fulfilled the wish that Robert Carson denied his sister by reading "a passage from her own hand which expresses in a remarkable way the strength, the simplicity, and the serenity that marked her character." The passage came from Rachel's letter to Dorothy when she recalled "that blue September sky, the sounds of wind in the spruces and surf on the rocks, the gulls busy with their foraging, alighting with deliberate grace, the distant views of Griffiths Head and Todd Point, today so clearly etched, though once half seen in swirling fog." Most of all Rachel remembered the monarchs and "the invisible force" that drew them each year to distant places from which they did not return.[1]

Those in the congregation that day must have sensed that through her writing she was linked to the past and future of the

environmental movement that she had done so much to ener-
gize. There was in their friend, they knew, a touch of the rebel-
lious spirit of Henry David Thoreau and John Muir, the conservation ethic of Theodore Roosevelt's chief forester Gifford
Pinchot, and the ecological ideas of zoologist Charles Elton.
Carson was especially drawn to Thoreau, who found in wilderness both an escape from the materialism of his day and a higher
spiritual truth and moral law. He associated nature with "absolute freedom and wildness, as contrasted with a freedom and
culture merely civil." A human being was best for Thoreau as "an
inhabitant or part and parcel of Nature rather than a member of
society." In addition to his romantic passions, Thoreau was, like
Carson, a dissenter who spent a night in jail rather than pay taxes
that supported a war against Mexico.

And again like Carson, Thoreau wedded his romanticism to
the sensibility of a scientist. He filled his library with botany texts
and collected specimens of plants because he "wanted to know
my neighbors" and "get a little closer to them." Some days he
walked as much as thirty miles to learn the exact moment when a
certain flower blossomed. What he wanted to understand was the
teeming diversity of the natural world around him. Over the last
ten years of his life, he turned himself into a naturalist in much
the same spirit that Carson worked as a field ecologist.[2]

Carson connected to the conservation tradition through her
respect for resource managers such as Gifford Pinchot, who created the system of national forests and wildlife preserves.
Supported by Theodore Roosevelt, Pinchot formed the Forest
Service in the early twentieth century to subject the nation's
woodlands to scientific management. He had seen the waste of

the nation's bountiful resources during the Industrial Revolution of the late nineteenth century, when profit trumped the interests of other potential users. Loggers, who clear-cut timber in the Adirondack and Catskill Mountains in upstate New York, for example, removed the forest cover that protected the water supply of New York City. Without trees and their roots to slow the runoff of melting snow and spring rains, soil washed into the Hudson River and its tributaries, making the water undrinkable without expensive filtration systems. Here was a hint of the human arrogance and myopia toward nature that Carson condemned in her later writing.

Conservationists of Pinchot's generation replaced that short-sighted approach with the concept of "multiple-resource usage." Rather than look at a forest as a single resource such as timber, scientific managers treated forests as a single resource with multiple uses—recreation, protection of watershed, wildlife preserve, as well as lumber. Scientific management allowed foresters to arbitrate among conflicting claims on woodlands. Judicious cutting practices turned timber into a renewable and more profitable resource, while leaving a healthy habitat for animals, recreation for hunters and hikers, and a barrier to erosion.[3]

Carson parted company with conservationists, however, when they became preoccupied with human uses of nature. The conservation ethic, as Pinchot practiced it, treated nature as a warehouse of resources for humans to develop. Such a view was too anthropocentric for Carson's taste. She identified more with Pinchot's contemporary John Muir, who devoted himself to preserving wilderness and other irreplaceable geographic areas. For Muir, a central measure of the value of nature was the benefit it conferred on human society. But while conservationists believed that benefit should be measured in economic terms, preservationists embraced the ideas of Thoreau and other Romantics who believed that nature provided intangible spiritual and aesthetic values to humans. Muir, the patron saint of preservation and a founder of the Sierra Club, became the chief propagandist for the ideal that nonmaterial values should have equal standing with economic ones. For Muir, a sea of wildflowers or a virgin forest left the viewer sensing the presence of a higher spirit. Yellowstone and Yosemite National Parks stood in Carson's time as monuments to the early successes of the preservationists.[4]

When Carson trained at Johns Hopkins, she became aware of an ecological perspective. In 1927, British zoologist Charles Elton published *Animal Ecology*, a book that became the basis for the "new ecology." In it, Elton defined a way of understanding nature that underlay all of Carson's writing. For Elton, and hence for Carson and other environmental scientists, communities of living things were at the heart of all natural systems. From Elton and the new ecology, Carson derived four key principles. First, she learned about the "food chain." Plants, with their ability to convert sunlight into food by means of photosynthesis, constitute the first link in the chain. As such, plants were the primary producers, those animals that lived off the plants were consumers, and those that fed off dead plants and animals were the decomposers. Second, in any natural community, the varieties and quantities of plants determined the array of consumers and decomposers. Removing any one species from the community threatened the survival of the whole.

Third, Elton taught Carson's generation that a relationship existed between the size of food resources and the species population in a food chain. Large mammals such as pandas require too much food to live off tiny plants or animals. Hence, pandas live where the bamboo they consume is abundant. Without the bamboo, pandas cannot survive because they cannot eat down or up the food chain. In most chains, small consumers turn food into larger forms eaten by larger consumers. Or, as Carson showed in her books on the ocean, the big fish eat the medium fish that eat the small fish that eat the plants. To make this system work, the small food sources must be the most fertile and the most abundant. Consumers at the top of the chain, such as sharks and whales, are far less numerous than the plankton at the base.

The fourth factor in natural communities, as Elton observed them, was the "niche." Each living thing, no matter where it fit in the food chain, filled a place, or what Elton thought of as an "occupation." It had a job to do or, more precisely, a food to eat. Every chain, for example, has creatures such as the vultures or the seashore crabs that live off the remains of other creatures. All the same, in no single community can two creatures fill the same niche. As he applied the principles of evolution, Elton believed that the competition for food was so intense that only one species could survive.

What made Elton's work so attractive to Carson was his notion of ecology as "scientific natural history." He placed as much emphasis on observations from fieldwork as he did on experiments in the laboratory. Conservation became for him an important outcome of applied ecology. He urged scientists to understand the dynamic of natural habitats so "we will be able to face conservation problems and understand what goes wrong in our artificially simplified farmlands and planted forests."[5]

Carson discovered in the new ecology another strain of thinking that contradicted the sense of wonder engrained in her when her mother introduced her to nature studies. A group of scientists led by the English botanist A. G. Tansley, was determined to banish from their field any notion of living organisms as communities. Ecologists such as Elton and Carson believed that communities in effect formed beings greater than the sum of their parts. Some implied that these communities possessed an organic intelligence or spirit of their own. For Tansley, this notion defied empirical proof. The whole of a community, he argued, could only be understood as the sum of its parts. The task for science then was not to speculate about this unknowable community intelligence or spirit but to find "the basic units of nature" and analyze these individual parts. Nature was exclusively material, and its behavior followed mechanical principles. In Tansley's view, true science limited itself to that which could be measured and analyzed.

To rid his science of any notion that communities in nature were somehow comparable to human societies or organisms, Tansley introduced the concept of the "ecosystem." Whatever unity existed among living things in a natural community derived from the exchange of energy and of the nutrients and chemicals essential to life. Where people such as Elton might distinguish between living and nonliving materials, Tansley saw them as intertwined. In his ecosystems, whether grand or small, complex or simple, all of nature, including nonliving rocks and gases, interacted as material parts of the larger whole. They were subject to the same laws of physics and chemistry that ruled the material world.[6]

Those who knew Carson understood her debt to the new ecology as well as to the traditions of conservation and preservation. Like conservationists, Carson marshaled the authority of science to make her case that pesticides threatened natural habitats and human health. She also shared with John Muir and Thoreau the conviction that humans needed to reconnect to the nonhuman world. She embraced Elton's idea of food chains, but she rejected Tansley's view that nature must be understood in narrowly material terms. To study living things exclusively in the laboratory and to reduce natural communities to their component parts blinded humans to nature's larger meaning. Carson instead believed in a realm beyond science, that "even with all our modern instruments for probing and sampling the deep ocean, no one now can say that we shall ever resolve the last, the ultimate mysteries of the sea." At the same time, the conservationist in her warned of "the possibility of the extinction of mankind by nuclear war" and of the "contamination of man's total environment with such substances of incredible potential harm" that they threatened "the very material of heredity upon which the shape of the future depends."[7]

No one knew better than those who gathered to honor Carson on that bright spring day how hard she struggled to warn against human abuse of the earth or the personal price she paid in waging her crusade. In memory of her heroic efforts and her inspiring prose, they initiated a process that one scholar compared to a "secular canonization," inspired by a vision of "Saint Rachel of the *Silent Spring*." Environmentalists soon elevated her to the pantheon of their saints. The Rachel Carson Trust for the Living

Environment she funded in her will became the Rachel Carson Council, with the goal of promoting "alternative, environmentally benign pest management strategies to encourage healthier, sustainable living." Her papers were stored in Yale's Beinecke Library, a temple to the printed word; the Rachel Carson National Wildlife Refuge preserved her favorite habitats along the Maine coast; and in Chicago, the Rachel Carson Scuba Corps dove on ecofriendly missions.[8]

Within a few years of Carson's death the environmental movement gained wide public support. Like Carson, it drew on the older traditions of the nineteenth-century Romantics, the twentieth-century conservationists and preservationists, and the new ecology. Her evocation of the mysteries of nature appealed to a growing community of dissenters, who in the 1960s resisted the idea that all that was worth knowing could be discovered through empirical observation. Some of these dissenters formed what came to be called the "counterculture." They sought to discover life's mysteries through ancient and Eastern religions, psychotropic drugs, and the culture of Native Americans. For them, astrology, Zen Buddhism, the earth goddess Gaia, and other metaphysical systems offered more promise than hard science for revealing a higher-order truth. Many joined rural communes to escape the materialism of modern America and bring themselves closer to nature. These communes constituted what Murray Bookchin, a Marxist critic of consumer society, called affinity groups and

Numerous cartoons captured Carson's legacy. (SOURCE: Cartoon by Conway, 1963. Courtesy of the Lear/Carson Collection, Connecticut College.)

social ecology. Such groups were united by their intention to live in ecologically responsible ways. They wore wool, cotton, and other natural fibers and grew what they ate without fertilizers, insecticides, or herbicides. Consumption did not go beyond the necessity of keeping mind and body healthy. In that way, spiritual goals replaced material ones. As human beings reduced their demands on the world's resources, they believed that nature would, as Carson hoped, return to health.[9]

Carson's critique of science run amok also found a responsive audience among dissenters who adopted a political analysis of society. A cluster of issues led them to protest against the current state of American society: racial inequality, poverty, the war in Vietnam, excessive materialism, or all of these issues. Carson helped focus their attention on the environment as well. They found in her attack on pesticides further proof of the evils of corporate capitalism. The desire for profit, they argued, more than any public good, encouraged chemical companies to market toxic materials that damaged the earth. Advertising masked the true cost to public health and the environment. Equally important, dissenters shared Carson's belief that "the people" had the right to influence decisions that affected their lives.

By the mid-1960s a series of events brought new urgency to Carson's warning that the destruction of nature threatened human survival. A study released in 1965 revealed that within the entire United States only one river near a major city (the St. Croix between Minnesota and Wisconsin) remained unpolluted. As the nearly all-black neighborhood of Watts erupted in a bloody riot in Los Angeles, a strike by sanitation workers nearly crippled New York City. Mountains of uncollected garbage stood as fetid symbols of society's wastefulness. Then, in November, a power failure blacked out seven northeast states and two Canadian provinces. The massive outage hit New York at the height of the rush hour, stranding hundreds of thousands of people in elevators and subways. These events exposed the fragile infrastructure on which the modern social order rested. People who might once have ignored Carson's discussion of "the web of life" in nature saw its counterpart in human society. Cities seemed to have their own ecosystems that now appeared vulnerable to cataclysmic disruption.

In January 1969, California learned firsthand about the damage an ecological disaster could do. For the affluent Californians who lived there, Santa Barbara was an earthly paradise fronted by white-sand beaches and panoramic ocean views. Unfortunately for its inhabitants, their modern paradise ran on oil. One of America's richer reserves of petroleum lay in the channel stretching between Santa Barbara and Los Angeles 90 miles to the south, where oil companies had drilled some 925 wells. Santa Barbara city officials worried about the possibilities of pollution, but oil companies frustrated state efforts to impose stricter regulations, while federal officials assured local leaders they had "nothing to fear."

How wrong they were. Beachcombers were aghast in late January to see that Union Oil's well A-21 was spewing a billow of gooey crude into the channel. Crews quickly capped the well, only to discover that the pressure had opened a fissure in the channel floor. Natural gas and crude seeped to the surface. "It looked like a massive, inflamed abscess bursting with reddish-brown pus," one observer remarked. Given the nature of ocean tides and currents, the disaster quickly spread. Within days, heavy oil tar covered five miles of Santa Barbara beaches. An oil slick extended some 800 miles south to San Diego. Along the blackened shoreline, birds lay dead and dying, most unable to even raise their oil-soaked feathers.

Few Santa Barbarans were surprised to learn that Union Oil was operating below state and federal standards when the well blew. Twice in the recent past the company had been fined for polluting California waters. Its president, Fred Hartley, dismissed the leak as "Mother Earth letting some oil come out." He later told a group of conservationists that he was "amazed at the publicity for the loss of a few birds." Such callous disregard for the environment confirmed growing fears that human carelessness threatened to destroy the ecosystems that supported life on earth. [10]

Then, in June 1969, the unthinkable happened. The Cuyahoga River in Cleveland, Ohio, long an industrial sewer and toxic waste dump, burst into flames. A river caught fire! Flames climbed as high as fifty feet until fireboats brought them under control. The irony prompted singer Randy Newman to croon, "Burn on, big river, burn on." *Time* magazine described the Cuyahoga as the river that "oozes rather than flows" and in which a person "does not drown but decays." Fire made it a national symbol of long-term patterns of human abuse of nature. [11]

These domestic disasters occurred against the backdrop of a seemingly unending war in Vietnam. And by 1969 Vietnam had become an environmental as well as a military battle. Critics of the war viewed it as one of the great ecological disasters of the modern age. As early as 1965, campus protesters targeted companies such as Dow Chemicals that produced napalm. When used as a bomb, napalm, a jellied gasoline, burned with such ferocious intensity that it melted the flesh off its victims. Less vicious, but more devastating, were herbicides such as Agent Orange. Frustrated by the thick jungle canopy that afforded cover to the Vietnamese insurgents, called "Vietcong," American generals made war on the land as well as its people. They sprayed large swaths of countryside with the defoliant Agent Orange to kill the vegetation and crops. Before the war ended, Agent Orange had laid bare an area roughly the size of Connecticut. The American troops who handled it and people exposed to it on the ground suffered unusually high rates of cancer and other diseases.[12]

To antiwar protestors, environmental destruction in Vietnam seemed to be an extension of the misguided effort to conquer nature. The terms "ecocide" and "ecotastrophe" echoed Carson's description of chemical sprays as "biocides." The assault on Vietnam's ecology seemed yet another demonstration of the destructiveness of the anthropocentric American way of life. Some protestors linked environmental destructiveness to corporate profiteering. A California group published a pamphlet entitled "Where There's Pollution, There's Profit." One radical environmentalist went so far as to argue that even at the height of the Cold War there was little difference between the United States and the Soviet Union: the "deterioration of the natural environment all around us is clearly the product of the nature of production and consumption . . . that today holds sway in technological society—American or Soviet." By 1969 many radicals saw ecological politics as another way to discredit capitalism and promote a revolutionary spirit.[13]

The growing radicalism of political and environmental protest worried Senator Gaylord Nelson, a progressive Democratic senator from Wisconsin. In September 1969, Nelson proposed a national teach-in as an effective way to inform Americans about current threats to the environment. He believed that the environmental crisis was "the most critical issue facing mankind," but he also saw the teach-in as a way to forge a movement free from the disruptive behavior of cultural and political radicals. Enthusiasm for Nelson's teach-in was so infectious that his office was soon deluged with requests for information on the event.

Nelson knew he needed someone to help turn his idea for "Earth Day" into reality. He found that person at the Harvard Law School. Denis Hayes embodied the divided sensibility of environmental activism. Hayes shared much of the radicals' conviction that government and business colluded to maintain an industrial order hostile to nature. "Ecology is concerned with the total system—not just the way it disposes its garbage," he told a press conference. He also believed that confrontational politics alienated mainstream Americans, whose sympathies were essential if the environmental movement were to succeed. "We didn't want to lose the 'silent majority' just because of style issues," Hayes explained. The way to proceed, he decided, was to transform protest into celebration.

Many of the more traditional environmentalists worried that the spirit of Earth Day would draw attention away from the conservation movement. The new environmental ethos was about living in harmony with nature and saving the earth's ecosystems. Traditional conservationists were more narrowly concerned with practical matters such as protecting natural resources or improving the national parks. "We cannot afford to let up on battles for old-fashioned wilderness areas, for more preservation of forests and streams and meadows and the earth's beautiful and wild places," a Sierra Club official argued. Instead, organizations should continue to focus on those strategies and goals that "the traditional movement has pioneered and knows best." Thus, some of the mainstream conservation and preservation organizations including the Sierra Club, Wilderness Society, and National Audubon Society steered clear of the first Earth Day demonstrations.[14]

Despite those reservations, on April 22, 1970, some 10 million Americans and millions more around the globe reclaimed city streets for pedestrians, planted trees, hiked, and in various ways protested the pollution of the environment. Television and newspaper editorials applauded the new environmentalism as a broad-based movement dedicated to stopping litter, cleaning the air and water, and preserving the wilderness. Earth Day drew much of its style and energy from the teach-ins, marches, and campus demonstrations of the 1960s. Many who gathered that day shared the idea of alternative lifestyles and the utopian visions that had inspired the romantics and preservationists of an earlier day. Yet, environmentalism did not briefly flourish and then die as upheavals of the 1960s gave way to the conservative backlash of the 1970s. A host of old and new environmental organizations fought to reduce pollution, advocated the rights of all living things, and rejected the assumption that humans had the right to control nature. No event could have done more to celebrate the ideals Rachel Carson bequeathed to the environmental movement.

So powerful was this new consciousness that in 1969 Congress began to enact an unprecedented body of environmental laws. With little opposition, it first adopted the National Environmental Policy Act (NEPA). The primary supporters of the NEPA had a rather narrow view of its purpose. They wanted to protect recreational uses of nature such as hunting, fishing, hiking, and camping while addressing issues affecting the quality of life: smog, water pollution, and waste dumps. Environmentalists turned it to far broader purposes, employing its provision that the government and its agencies must report on the environmental impact of their activities as a reason to monitor everything from federal road building to draining wetlands. Congress then created the Council on Environmental Quality to ensure that federal agencies complied with the new law. The need to hold public hearings and report on environmental impacts of such activities put teeth into the NEPA.[15]

Congress followed by creating the Occupational Safety and Health Administration (OSHA) to ensure workers were not exposed to hazards that included toxic chemicals and dangerous substances such as asbestos. Additional legislation guaranteed clean air and water, limited insecticides, and protection for endangered species. Seeing that the prevailing political winds now blew with the environmentalists, Richard Nixon set his sails in their direction. In 1970, Nixon brought together numerous federal agencies to create the Environmental Protection Agency (EPA). As the first director, William Ruckelshaus recognized that his agency faced the daunting task of forming "a cohesive functioning, integrated entity out of fifteen agencies and parts of agencies throughout the federal government."[16]

Determined to establish the authority of the EPA over the environmental agenda, Ruckelshaus treated DDT as an important test case. In passing the NEPA, Congress shifted authority over pesticide regulation from the Department of Agriculture to EPA, even though, in the years following the publication of *Silent Spring*, pressure from industry lobbyists stymied congressional efforts to ban its use. Ruckelshaus at first agreed with the chemical industry that DDT posed no imminent hazard to public health. All the same, he faced almost irresistible pressure from environmental groups to phase it out.[17]

After a federal court canceled all uses of DDT, a coalition of chemical manufacturers demanded a hearing to appeal their case. From August 1971 until March 1972, DDT was almost literally on trial. Expert witnesses and attorneys for both the industry and their environmentalist adversaries produced some 9000 pages of testimony before a hearing examiner. Twenty-seven manufacturers, the USDA, and several small DDT agricultural users defended the insecticide; EPA, the Environmental Defense Fund, the Sierra Club, the National Audubon Society, and the Western Michigan Environmental Action Council argued against it. (The Audubon Society helped to finance this effort with money from a fund Carson had provided in her will.)

Each side presented its evidence and then faced hostile cross-examination from opposing attorneys. The burden of proof lay with the industry and its allies, who had to establish that DDT was not harmful to animals, to consumers, or to those who applied it. Most contentious was the question of whether DDT caused cancer or other human health problems. A lesser issue arose about whether DDT was needed to control disease-bearing insects. The surgeon general testified that in the past twenty years no situation had arisen in which DDT played a crucial role in containing insect-borne disease. Malaria, for example, once a major killer in the United States, had been eradicated. But since DDT was still manufactured in the United States, its defenders argued that a ban might have a devastating effect on those countries that imported it to fight diseases.

In the end, the evidence about human health consequences was hardly clear-cut. Some scientists conducted experiments with DDT that produced tumors in lab animals; other scientists claimed to find no adverse effects in animals exposed to the chemical. The USDA tried but largely failed to discredit evidence showing that DDT contamination harmed domestic animals and wildlife. Agriculture Department scientists instead argued that DDT was still essential in some situations but should soon be replaced.

The EPA and its allies also attempted to show that DDT imposed a burden on the environment. In making that case, they had powerful support. After hearings held in Wisconsin in 1969 in which many of the same experts had testified, the hearing examiner ruled that DDT existed in all levels of the food chain and that "the only valid permissible inference is that DDT in small dosages has a harmful effect on the mammalian nervous system." Wisconsin determined DDT was a pollutant and banned its use. Two federal commissions, including one created by the hearing examiner Edward Sweeney, also established that, despite diminishing use, DDT had spread through world ecosystems. Because it concentrated in living organisms, it posed a threat to at least some species. Even if no evidence established a clear threat to human health, both bodies recommended nonetheless that DDT be phased out within a few years.

Industry lawyers subjected EPA witnesses to hostile cross-examination but so at times did hearing examiner Sweeney. His conduct struck EPA attorneys as so biased that they began to question his impartiality and competence. Some witnesses even refused to testify until he agreed to treat them with respect. Thus, few were surprised when, in the face of substantial testimony against DDT, Sweeney decided the case in favor of the industry. He argued that simply because evidence suggested DDT was carcinogenic in lab animals, the same conclusion could not be applied to humans. As a consequence, those who testified that DDT did not harm humans made a more persuasive case. Against any possible indications that DDT posed risks must be weighed "the well-documented proof of the benefits that DDT has bestowed on mankind."[18]

Sweeney did not have the last word in determining federal policy in this matter. Final authority rested with William Ruckelshaus and the EPA. For Ruckelshaus the decision had implications far beyond the basic question of whether DDT posed risks to humans or animals. On one side Ruckelshaus faced "the environmental movement, pushing very hard to get emissions down no matter where they were—air, water, no matter what—almost regardless of the seriousness of the emissions." On the other side stood industry "pushing just as hard in the other direction and trying to stop all that stuff, again almost regardless of the seriousness of the problem." The decision about DDT would be a critical indicator of how vigorously the EPA planned to enforce new environmental regulations.[19]

Ruckelshaus could not ignore the powerful scientific evidence against DDT. Nor could he forget that *Silent Spring* had persuaded the general public that DDT and other hydrocarbon-based pesticides posed a danger to humans and nature. In June 1972, Ruckelshaus rejected Sweeney's ruling and banned the spraying of DDT on crops. The pesticide could still be employed for public health emergencies and manufactured for export, but its widespread use in the United States would end in six months. In reaching his decision, Ruckelshaus concluded that evidence "compellingly demonstrates" that DDT threatened fish and wildlife and was a "potential carcinogen" in humans. Not only did this decision give the EPA broad regulatory authority, it also slew the dragon Carson had attacked. Prior to the publication of *Silent Spring*, the USDA registered pesticides based on evidence supplied almost exclusively by manufacturers. The public had no voice in those decisions. Now, with the EPA in control, people such as Carson's friends Olga Huckins from Massachusetts and Marjorie Spock from Long Island would have a role in regulating new chemicals that they believed threatened their health or their property. In banning DDT, Ruckelshaus recognized the power of ordinary people.[20]

This victory was hardly complete. Conservative opponents of environmental regulation argued then and after that, while Sweeney was competent to reach a proper decision, Ruckelshaus was not. They charged that for Ruckelshaus, but not for Sweeney, politics outweighed science. Chemical companies continued to press their case, even after an appeals court ruled against them in 1973. Agricultural interests asked for an emergency waiver so that they could use DDT against the pea leaf weevil in 1973 and the fir tussock moth that attacked Douglas firs in the Pacific Northwest in 1974. More important, powerful members of Congress resented the authority the EPA claimed over pesticide regulation and occasionally threatened to pass legislation overriding its rulings.

With the election of Ronald Reagan in 1980, conservatives launched a powerful attack against the political and cultural legacy of the 1960s. One of their primary targets was the environmental movement and the regulatory apparatus that it supported. Economic conservatives, who resented government regulations, and libertarians, eager to limit the size and power of government, chafed at the host of laws and agencies created to improve the quality of air and water, save wilderness, protect endangered species, and limit pollution. Many business leaders charged that environmental red tape hamstrung the nation's ability to compete against foreign manufacturers; libertarians argued that regulation denied private citizens their right to exploit public resources.[21]

Bridging those views, Reagan's secretary of the interior, James Watt, maintained "failure to know our potential, to inventory our resources, intentionally forbidding proper access to needed resources, limits this nation, dooms us to shortages and damages our right as a people to dream heroic dreams." With missionary zeal, Watt set about reordering the nation's priorities: opening vast areas to mining and oil drilling, refusing to enforce environmental rules and regulations, turning national parks over to private concessionaires, and selling off public lands. In time, Watt's zeal turned him into a political liability and forced his resignation in 1983. He was not, however, a lone voice crying against the wilderness but rather a more outspoken representative of what came to be known as the "Sagebrush Rebellion" and, more generally, the "Wise Use" movement. These conservative rebels were determined to free property owners from environmental regulation and argued that transferring government-controlled resources into private hands would promote economic growth.[22]

To their dismay, Reagan's Wise Use allies discovered that the public did not want to abandon environmental rules and favored the cleanup of the nation's air and water. The Reagan revolt against environmentalism proved stillborn. Efforts to dismantle the EPA did not simply die, however, nor did the Wise Use rebels fold their tents and head for high ground. In 1994 they were back in action as Congressman Newt Gingrich of Georgia helped Republicans win control of both the House of Representatives and the Senate. The centerpiece of their broad agenda to limit government and restore "family values" was a plan to reduce environmental regulation.

The EPA once again lay in their sights. Congressman Tom DeLay of Texas called their plan for deregulation "Project Relief." As a former exterminator, DeLay knew something of the restrictions imposed by environmental rules. He boldly announced to the press, "You've got to understand, we are ideologues. We have an agenda. We have a philosophy. I want to repeal the Clean Air Act." One moderate Republican senator, dismayed at the boldness of DeLay's plan to revise environmental law, called it "terrible legislation. When all the artichoke leaves are peeled away, they are out for the Clean Air Act, the Clean Water Act, the Endangered Species Act; that is what they are gunning for."[23]

Having set out to destroy the environmental movement, these conservatives soon trained their fire on Rachel Carson. How better to discredit the movement than by tarnishing the reputation of its patron saint? In many ways, their tactics differed little from those used in the 1960s by critics of *Silent Spring*. Carson, they charged, had practiced bad science and thereby misrepresented the value of DDT. More than that, the reverence for nature she encouraged created formidable roadblocks to the rapid exploitation of natural resources they advocated.

Political scientist Charles Rubin launched an early attack to discredit Carson's science, though his primary target was the "green crusade" to protect the environment and not simply Carson. Rubin agreed with Paul Brooks's assessment that *Silent Spring* was "one of those rare books that change the course of history . . . by altering the direction of man's thinking." Given what he saw as the shoddiness of her research, how then to account for its enormous significance, Rubin wondered. Was it her literary skill? Rubin thought not since he believed nature writing was a well-established genre before Carson began to publish. Was it her warning against pesticides? No, others had done so before her and nothing in her research for *Silent Spring* was original. Then suddenly, Rubin revealed his purpose. "Was it her fanatical attacks on DDT and modern chemical technology that set the tone for the subsequent excesses of environmental fear-mongering?" he asked, inviting no answer.[24]

Rubin did make a persuasive case that Carson was not always neutral in her use of sources and that she was sometimes driven by moral fervor more than by scientific evidence. Indeed, her use of evidence was selective, and she made no attempt to catalogue the benefits of pesticides, as critics such as Rubin insisted she should have. But there was no need to. The chemical giants of the 1950s required little defending. They had at their disposal vast public relations and advertising resources and spent millions extolling the virtues of pesticides without ever acknowledging their toxicity. Carson provided information neither the companies nor the government had ever made public but that she believed people had both a need and a right to know. Despite his determination to discredit Carson, Rubin concluded rather meekly that *Silent Spring* was "not a fully coherent or completely worked out line of argument." That was a point Carson would have readily conceded. She worried continually about the accuracy of her facts and her sense that the subject was far too complex for a single author to master, let alone one battling to stay alive.

Rubin's tepid attack against Carson suggests he had a bigger target in mind. Writing in 1995, shortly after the fall of the Berlin Wall and the collapse of Communism, Rubin believed that Carson's disciples in the environmental movement, those he dismissed as "utopian reformers," posed a new threat to freedom and liberty because he suspected that "as 'red' totalitarianism declines, the aspirations of our radical reformers may become increasingly 'green.'" In that way, many conservatives believed that environmentalists now replaced domestic Communists as the enemy Americans should fear.[25]

Other detractors of Carson claimed that her arguments were not based on science but on faith. They described environmentalism, as Carson and her followers envisioned it, as a pseudoreligion. "Green worshippers can keep their religion," snorted one critic of ecology, while another described it as a "new religion, a new paganism that worships trees and sacrifices people." Physician and novelist Michael Crichton called environmentalism "one of the most powerful religions in the Western World." In the spirit of Judeo-Christian beliefs, Crichton suggested, environmentalism begins with an Eden, "a state of grace and unity with nature; there's a fall from grace into a state of pollution as a result of eating from the tree of knowledge, and as a result of our actions there is a judgment day coming for all of us. We are all energy sinners, doomed to die, unless we seek salvation." In this environmental religion, salvation comes as "sustainability." Crichton dismissed the "green" worldview as romantic claptrap. "People who live in nature," he assured his audience, "are not romantic at all."[26]

Some antienvironmental critics moved far beyond accusing Carson of bad science and misguided beliefs. They accused her of murder. On a blog site in 2003, one outraged writer claimed that a child died every 15 seconds, 3 million people succumbed each year, and since 1972 100 million people were lost because of a pandemic sweeping the globe. "These deaths," the author fumed, "can be laid at the doorstep of author Rachel Carson." How had Carson done this horrid deed? "Her 1962 book *Silent Spring* detailed the alleged 'dangers' of the pesticide DDT, which had practically eliminated malaria." Without that "cheap, safe, and effective" weapon to control insect-borne disease, "millions of people—mostly poor Africans—have died due to the environmentalist dogma propounded by Carson's book." Writing in the *New York Times* in April 2004, Tina Rosenberg put the issue only a bit more soberly: "DDT killed bald eagles because of its persistence in the environment. *Silent Spring* is now killing African children because of its persistence in the public mind." Michael Crichton abandoned derision for moral outrage. "Banning DDT is one of the most disgraceful episodes in the Twentieth Century history of America," he commented. "We know better, and we did it anyway, and we let people around the world die and we don't give a damn."[27]

These latter-day Carson critics marshaled some disturbing evidence. According to Rosenberg, health officials estimated that malaria killed 2 million people a year; the large majority of those who died were children under five living in Africa. Until the recent explosion of the AIDS epidemic, malaria was Africa's leading killer, taking the lives of one in twenty children and leaving countless millions of others brain-damaged. Worldwide, 300 to 500 million people contract malaria each year. Beyond the loss of life and untold suffering it caused, malaria shrank the economies of some of the world's poorest countries by some 20 percent over fifteen years. What so disturbed Crichton, Rosenberg, and others was not only that people in the wealthy Western countries largely ignored this pandemic but also that a cheap and effective way to curb it existed. The solution, Rosenberg asserted, "lasts twice as long as the alternatives. It repels mosquitoes in addition to killing them, which delays the onset of pesticide-resistance. It costs a quarter as much as the next cheapest insecticide." The answer, Rosenberg assured her readers, was DDT. Semiannual spraying of the interior of huts in South Africa with DDT greatly reduced malaria and the related health costs.[28]

Even though the EPA banned DDT in 1972, it did not ban its manufacture nor did international law prevent other countries from using it. What then stopped the poor nations of the world from spraying DDT to curb the malarial scourge? The simple answer offered by these critics was "Rachel Carson." The prejudice against DDT that Carson aroused in *Silent Spring* made it difficult for health officials to press for its use. In addition to the United States, most other developed nations banned DDT. By 2005 this situation put the third world in a double bind. Health organizations that might assist countries in the battle against malaria received their funding from the wealthier nations where prejudice against DDT was strongest. One foreign aid administrator admitted that her agency would not finance DDT because "you'd have to explain to everyone why this is really O.K. and safe every time you do it, so you go with the alternative that everyone is comfortable with." Most critics of the DDT ban argued those alternatives such as bed netting infused with insecticide were almost always more expensive and less effective.[29]

Equally problematic for DDT advocates, many developing world agricultural producers also supported the DDT ban. The United States and Europe generally barred any imported crops containing traces of the pesticide. Beyond that, the chemical companies that once attacked Carson and vigorously defended DDT no longer did so because DDT was no longer under patent, which meant that any company could now manufacture it. Hence, chemical companies found it more profitable to sell other insecticides. One economist observed, "Clearly, they'd like to see DDT banned—it cuts into their markets." As a consequence, while much of Africa lacked the means it needed to fight malaria, the continent's children continued to die.[30]

This case would offer a withering indictment of Carson and "her coterie of admirers" if it held up to close examination. Is it true, for example, that 300 to 500 million people a year become ill with malaria? That number seems unreasonably high. If it were accurate, then within as few as thirteen years malaria would infect all of the world's 6.5 billion people. Clearly, that is not the case. What then does that number actually mean? Not, in fact, that 300 to 500 million people contract malaria each year but that 300 to 500 million people show symptoms of the disease. Many of those "incidence[s] of clinical disease episodes" occurred in people who had been infected for many years and were not, in fact, new cases.

And what of the 1.1 to 2 million people who die from malaria, most of them African children under the age of five? Would the wider use of DDT reduce their level of mortality? Surely, such a number would justify a worldwide effort to contain the disease, even if that meant a significant increase in the use of DDT. But statistics collected by the World Health Organization might give another perspective on that death rate. As frightening as malaria may be, it ranks only fourth as a cause of infant mortality. Respiratory infections and diarrhea claim over twice as many lives as malaria, while neonatal conditions such as preterm birth and low birth weight kill as many 10.6 million infants annually. AIDs is another deadly disease, but its victims die less often in infancy. Such comparisons suggest that if those who condemned Carson and *Silent Spring* truly cared about the children of Africa, they might move beyond their preoccupation with malaria and DDT and propose alternative solutions to the AIDS crisis, water pollution, malnutrition, and infectious diseases that are the continent's greatest killers. Unlike malaria and DDT, no magic bullet can cure the most serious of these ills. Improved maternal care, adequate diet, better sanitation, and clean drinking water would all do more to cut infant mortality than spraying with DDT; but those solutions are more expensive and difficult to achieve. The ills afflicting Africa require a more comprehensive plan than DDT's advocates have supported.

And is it true that DDT poses no significant risk to humans? Among her critics, that was a recurring argument. One advocate claimed "if you use DDT properly, it has a record of safety and effectiveness for humans that is really unmatched." By demonstrating that DDT is in fact harmless, critics enhanced their claim that Carson's work was based on bad science, unwarranted assumptions, and "fear-mongering." Yet, two highly respected researchers who studied the impact of DDT exposure on preterm births and the length of time mothers nursed estimated from their data what might happen if Africans resorted to wider indoor spraying. They found, for example, that among women in Mexico and North Carolina, those with higher concentrations of DDT lactated for shorter periods. In Africa, where food is scarce, mothers breast-feed for an average of eighteen months. If lactation there fell to the levels found in Mexico and North Carolina, infant mortality might reach a level that wiped out any benefit from spraying DDT. Those who have spoken with such moral passion about the virtues of DDT may well have been recommending a cure that was more dangerous than the malaria they were trying to eradicate.[31]

In the early twenty-first century no more than in the 1960s were critics properly representing what Rachel Carson was saying. Carson would have been among the first to support limited applications of DDT in order to save lives. In *Silent Spring,* she never spoke against responsible use of pesticides. Rather, she urged that the methods of insect control "be such that they do not destroy us along with the insects." What she did condemn were uncritical and often untested claims that these chemicals were harmless to humans and other living things. She further decried the anthropocentric point of view that saw humans as somehow separate from and not responsible for their impact on nature. For Carson, life was a wondrous mystery to behold, and she asked her fellow beings to be aware "that we are dealing with life—living populations and all their pressures and counter-pressures, their surges and recessions."

By contrast, Carson's critics remain wedded to the notion that humans can and should control nature. They resent the limits placed on economic growth by those seeking to protect the environment. As one frustrated critic of Carson remarked, "Those worried about the arrogance of playing God should realize that we have forged an instrument of salvation, and we choose to hide it under our robes." To that charge the gentle Rachel Carson might have said that those who saturate the fabric of life with chemicals bring to their campaign "no humility before the forces with which they tamper." The subversive Rachel Carson might have suggested that such arguments show a "lack of prudent concern for the integrity of the natural world that supports all life."[32]

Above all, the gentle subversive advocated a holistic ethos that located humans within the "web of life." She urged parents to nurture in their children "a sense of wonder so indestructible that it would last throughout life, as an unfailing antidote to boredom and disenchantments of later years, the sterile preoccupation with things that are artificial, the alienation from the sources of our strength." In the 1920s, she witnessed that alienation in her classmates at the Pennsylvania College for Women. As a writer, she gave her readers a sense of the awe that fired her own imagination. Had she lived only a few years more, she would have seen how the struggle with boredom, alienation, and artificiality inspired much of the youthful unrest of the 1960s out of which sprang the modern environmental movement. That movement embodied her hopes for reconciliation between humans and nature. And had she survived into the twenty-first century, she likely would have picked up her pen again to condemn the critics of environmentalism as proponents of a crusade to revive "the Neanderthal age of biology and philosophy, when it was supposed that nature exists for the convenience of man."[33]

Notes

1. The events of the funeral are covered in Lear, *Rachel Carson*, pp. 480–483.

2. On Thoreau, see Worster, *Nature's Economy*, pp. 57–76, and Nash, *Wilderness and the American Mind*, pp. 84–95.

3. Samuel P. Hays, *Conservation and the Gospel of Efficiency* (Cambridge, Harvard University Press, 1959), pp. 1–4. Some important contrary ideas are in "The Value of a Varmint," in Worster, *Nature's Economy*, pp. 258–290.

4. On Muir and the preservationists, see Nash, *Wilderness and the American Mind*, pp. 122–140; on the clash between Muir and Pinchot over the Hetch Hetchy Valley, Ibid., pp. 161–181. There is little doubt that Carson would have stood with Muir in the battle over the valley.

5. The link between Carson and Elton is developed in Frederick R. Davis, "'Like a Keen North Wind': How Charles Elton influenced *Silent Spring*." Paper delivered for the American Society for Environmental History, St. Paul, Minnesota, April 1, 2006; for a cogent analysis of Elton's contributions to ecological thought, see Worster, *Nature's Economy*, pp. 294–301.

6. Worster, *Nature's Economy*, pp. 301–304.

7. Ibid., p. 349; Carson, *The Sea Around Us*, p. 196.

8. Howarth, "Turning the Tide," p. 42.

9. For an analysis of the modern environmental movement, see Gottlieb, *Forcing the Spring,* pp. 81–116, and Philip Shabecoff, *A Fierce Green Fire: The American Environmental Movement* (New York, 1993), pp. 77–185. Another source is Lytle, *America's Uncivil Wars,* pp. 323–333.

10. Lytle, *America's Uncivil Wars,* p. 327.

11. *Time,* August 1, 1969.

12. A good source on this dark chapter is Fred A. Wilcox, *Waiting for an Army to Die: The Tragedy of Agent Orange* (New York, Seven Locks Press,1983).

13. Lytle, *America's Uncivil Wars,* p. 329.

14. On the politics of Earth Day, see Gottlieb, *Forcing the Spring,* pp. 105–114.

15. On NEPA, see Ibid., pp. 124–125, and Dunlap, *DDT,* pp. 209–210.

16. Ibid., p. 210.

17. Ibid., p. 211.

18. Ibid., pp. 211–230.

19. Shabecoff, *A Fierce Green Fire,* pp. 130–131.

20. Dunlap, *DDT,* pp. 231–234.

21. A good source for the growing antienvironmental movement is Helvarg, *The War Against the Greens.*

22. Ibid., pp. 8–14; Shabecoff, *A Fierce Green Fire,* pp. 208–209.

23. Jacob Hacker and Paul Pierson, *Off Center: The Republican Revolt and the Erosion of American Democracy* (New Haven, Yale University Press, 2005), p. 142.

24. Rubin, *The Green Crusade,* pp. 30–31.

25. Ibid., pp. 31–52; see also the jacket liner notes for the green menace quote.

26. Thomas Dunlap, *Faith in Nature: Environmentalism as Religious Quest* (Seattle, 2004), p. 5; see also Michael Crichton, "Environmental Extremism: Remarks to the Commonwealth Club of San Francisco," *Looking Forward,* Jan/Feb 2004, pp. 12–13.

27. Lisa Makson, www.frontpagemag.com, July 31, 2003; Tina Rosenberg, "What the World Needs Now Is DDT," *The New York Times Magazine,* April 11, 2004; Crichton, "Environmental Extremism," p. 14.

28. Rosenberg, "What the World Needs Now."
29. Ibid.
30. Ibid.
31. This reconstruction of Carson's defense depends heavily on Karaim, "Not So Fast with the DDT," pp. 53–59.
32. Carson, *Silent Spring*, pp. 297, 13.
33. Ibid., p. 297; Carson, *The Sense of Wonder*, p. 43.

AFTERWORD

BEFORE I READ *SILENT SPRING* I HAD NOT THOUGHT MUCH about environmental issues, even though Buffalo, where I grew up in the 1950s, was an ecological disaster in the making. Its steel factories belched a rust-colored smoke, and nearby Lake Erie nearly died from pollution. When the Dutch elm blight threatened the city's stately trees, local authorities made a futile attempt to save them by spraying with DDT. Reading Rachel Carson years later forced me to see a connection between the declining quality of the environment around Buffalo and the way Americans lived in the postwar era.

I lost track of Carson and the controversy she provoked as I headed off to college in 1962. The ideas she raised in her book were simply too new to hold the attention of a city boy for long. Besides, a multitude of issues confronted college students in the early 1960s. Racism, imperialism, consumerism, and sexism all engaged us. Not until over twelve years later did I rediscover Carson, when I moved to the mid-Hudson Valley region to begin teaching at Bard College. By then, the physical evidence of the degradation of nature was too pervasive to ignore, even in the bucolic setting around Bard celebrated by landscape painters of the nineteenth century. Suburban sprawl crept into the mid-Hudson area from Albany to the north and from New York City to the south. The majestic Hudson River that flowed by our town was too polluted for either swimming or fishing. From the Ohio Valley, electric utilities pumped smoke plumes into the air that fell as acid rain on our local ponds and forests. As my concern grew, I introduced a course in environmental history and with other Bard faculty and students created an environmental studies program.

In those days, conservation and preservation dominated the teaching of environmental history. I taught my students about the efforts in the late nineteenth and early twentieth centuries of conservationists to curb the waste of natural resources and of preservationists to protect unique landscapes such as the Yosemite Valley from commercial despoilation. The New Deal's Tennessee Valley Authority, or TVA, seemed an enlightened example of how humans could maximize their use of natural resources and pro- mote democracy, while restoring the health of one of the most ravaged areas in the United States. This approach quickly struck me as inadequate. The enlightened TVA of the 1930s was by the 1970s the nation's single largest polluter and a major source of the acid rain damaging northeastern forests and lakes.

At the same time, historians of the Trans-Mississippi West and the environment began to challenge some of the most widely held myths in American history. Where movies, fiction, and traditional historians celebrated the western march of the American people as a triumph of "civilization over savagery" and even good over evil, the new western history told a story of greed, environmental pillage, and the tragic destruction of indigenous peoples. In 1972, Alfred Crosby, Jr., demonstrated the possibility of an eco-logical approach to colonial history in his pathbreaking book *The Columbian Exchange: Biological and Cultural Consequences of 1492*. Crosby made a persuasive case that Spanish diseases, ani-mals, and plants had done more to destroy the native peoples than the Conquistadors and their muskets. At the same time, he argued that New World plants such as the potato revolutionized European food production. The history of European settlement of the New World was also an ecological story about the transfer of flora and fauna.

By the early 1980s environmental history had become a recog-nized field. Donald Worster gave it a frame and a past. In *Nature's Economy: A History of Ecological Ideas* (1977), he revealed that the science of ecology had been shaping human perceptions of the natural world for much longer than historians or scientists had previously understood. The term "ecology" first came into usage in 1866, but the ideas that informed it existed over a hun-dred years earlier. Eighteenth-century naturalists already perceived that all organisms, plant and animal, were part of "an interacting whole," or what they referred to as the "economy of nature." Over the next three hundred years that perception took scientists and nature lovers down many roads: Charles Darwin on the *Beagle*, Henry David Thoreau wandering the woods at Walden Pond, Frederic Clements surveying the windswept Nebraska grasslands, and Rachel Carson gathering samples from the rocky ledges along the Maine shore. Worster suggested that the ideas of ecol-ogy evolved through numerous phases between the eighteenth and twenty-first centuries, while maintaining elements from each of them. He likened the field to a man "who has lived many lives and forgotten none of them."[1]

Once historians had a sense of the science of ecology, they could use its theories to explore the past. In *Changes in the Land* (1983) William Cronon applied ecological analysis to unravel the process of the European settlement of colonial New England. Historians have wondered, for example, what New England forests might have looked like before the arrival of European settlers. From "tree rings, charcoal deposits, rotting trunks, and over-turned stumps," ecologists were able to recreate a proximate history of some New England woods. Fossilized pollen found in bogs and pond sediment provided another kind of evidence from which ecologists could determine the species composition of some ancient forests.[2]

Cronon argued that historians could not claim to explain human institutions without understanding the ecology that shaped them. The reverse was also true. Cronon and other environmental historians demonstrated that culture imposes ecological order on the land. What, after all, explains the stone walls that hikers frequently stumble upon in many of New England's forests? Once upon a time, the region's farmers cleared the forests to make a place for their farms. The boulders they pulled from the rocky soil made substantial fences for their fields. But over time, as New Englanders moved into towns and cities to make their livings, forests surrounded the walls as they reclaimed the land.

That shift from rural to urban took place as New Englanders began to create a market-oriented economy. As Cronon observed, this shift had profound consequences for the ecology of the region. In *Ecological Revolutions: Nature, Gender, and Science in New England* (1989), Carolyn Merchant brought a feminist perspective to this ecological story. The revolution that occurred, Merchant explained, had as much to do with reproduction as with production: "When reproductive patterns are altered, as in population growth or changes in property inheritance, production is affected." The reverse is also true, for as she added, "Conversely, when production changes as in the addition or depletion of resources or in technological or social innovation, social reproduction and biological reproduction are altered." So, for example, as European settlers created markets for fur-bearing animals, they disrupted the ecology upon which native peoples subsisted. New England farmers began to cut down the forests and wall off the lands on which the Native Americans once hunted. In time, as their traditional food resources disappeared, the Native Americans could not maintain population levels sufficient for their way of life to survive. Of course, European diseases had an equally devastating impact.

No more than the Native Americans could the Euro-Americans escape the consequences of ecological change. In the seventeenth century, as Merchant explained, because labor was scarce, colonial families needed large numbers of children to maintain their subsistence agricultural economy. Over time, however, land lost its fertility and farms became ever smaller as fathers divided their holdings among their sons. Many farms were no longer sufficient to support a family. By the late eighteenth century, when New England stood on the verge of a new market-oriented industrial economy, the region possessed a large pool of landless sons and unmarried daughters willing to work for wages. Women who had once been involved in both production and reproduction of the domestic economy now saw those roles separated. Production of such household goods as textiles and shoes became a public business, while reproduction remained domestic. Merchant concluded, "ecological revolutions are generated through tensions and interactions between production and ecology and between production and reproduction."[3]

In seeking to deepen my students' understanding of the concepts that framed the work of environmental historians, I turned to Aldo Leopold and Rachel Carson. In *A Sand County Almanac,* Leopold introduced the concept of "the land ethic" and urged humans to rethink their relationship to the earth, to become custodians and not exploiters. In *Silent Spring,* Rachel Carson warned of the danger of pesticides but also wrote about food chains, the balance of nature, and the interdependence of living things. From those two writers, my students learned some of the basic principles of ecology.

Having adopted *Silent Spring* in my course, I was delighted with the results. My students found Carson's prose engaging and her ideas compelling. The politically preoccupied among them thought that Carson offered vital evidence that self-aggrandizing corporations such as those that produced pesticides controlled the American political economy. Environmental studies majors tended to focus on Carson's ecological ideas and understood that she advocated a radical transformation of the consumerist values that sustained the American economy. Carson urged humans to reject the idea of controlling nature and to live humbly with the earth and its creatures.

More intriguing to me, my students found a political message I missed when I first read *Silent Spring*. Carson did not simply charge that the chemical industry and government scientists turned the public into unwitting guinea pigs as they conducted a war against insects and weeds. She also argued that the people had a right to know about experiments conducted in ways that put them and the future of their species at risk. She made no secret of her contempt for the men in white lab coats who hid the irresponsibility of their actions behind the prestige of their professions and the authority of the government. Carson advocated "power to the people" before radicals popularized the phrase in the 1960s

I finally realized that Rachel Carson, so genteel and proper in her personal life, was a subversive. She encouraged an entire generation of Americans to rethink fundamental values defining the relationship between human beings and nature by shifting from an anthropocentric notion of earth to a biocentric worldview in which people coexisted with nature and not over it. Only then could they take responsibility for the wanton destruction of nature done by others on their behalf and in that way reduce the growing threat to life itself.

Carson offered a scathing critique of corporate irresponsibility, misguided science, and government complicity in what amounted to a pollution scandal. She also taught my generation to appreciate ecology or, what in the early 1960s biologist Paul Sears called, "the subversive science." Ecology was subversive because it put nature rather than humans at the center of a living world in which everything is connected to everything else. And unlike most Americans in the 1960s, ecologists did not believe that science offered all the answers to the mysteries of life, nor could technical and scientifically engineered quick fixes solve the growing crisis of the natural world. Rather, they agreed with ecologist Paul Shepard, who joined Carson in calling for "an element of humility which is foreign to our thought, which moves us to silent wonder and glad affirmation." [4]

Of course, ecologists such as Sears, Shepard, and Carson never thought of themselves as subversives; they thought of themselves as disinterested scientists educating the public. It was their critics who pinned the label on them. But in the conformist culture of the postwar era, anyone who dissented against the orthodoxies of the day risked being dismissed as a crank or condemned as a traitor. If controversy was the inevitable price of bringing the nation to its senses before pollution poisoned the earth, it was a price these ecologists were willing to pay. Having been called subversives for expressing their beliefs, they wore the label proudly.

No more than she was subversive was Rachel Carson "gentle" in the way the popular media often represented her. From the age of fourteen, when she tried to negotiate terms with Author's Press, Carson showed a determined, even tough, business sense. Certainly, she demonstrated practical skills and managerial ability in supporting her family and in her work directing publications for the Fish and Wildlife Service. Her budget director there, John Ady, must have learned to respect her tenacity during their negotiations because, as often as not, he either compromised or yielded to her demands. Her research assistant, Bette Haney, recalled an incident when a community leader called to object to her views on a local pesticide-spraying program. He called her an "alarmist," Haney recalled, "and added patronizingly, when she began to cite case histories to him, she must not believe everything she read in the newspapers." His dismissive comments outraged Carson, as did his facile use of the uninformed ideas she fought to discredit. As Haney noted, "no one, seeing her attitude that afternoon, could have called her gentle."[5]

There were, all the same, recurring themes in Carson's writing that reinforced the image of her as a "gentle" woman. Carson grew up in a generation that became increasingly outspoken against cruelty toward animals, whether as laboratory guinea pigs, captives in circuses and zoos, maltreated pets, or the prey of hunters. This concern led in the late nineteenth century to the formation of the Society for Prevention of Cruelty to Animals. Carson expressed her sympathy for the movement in several ways. One was her respect in both word and deed for the sanctity of life. She celebrated the creatures of the earth for their beauty and the ingenious ways they adapted to the environment. Her friends frequently remarked that she insisted on returning all research specimens she collected to their habitats. By contrast, hunting perturbed her for its cruelty to individual animals but also because it threatened some species with extinction. Among the most vivid and disturbing images in *Silent Spring* are those describing the symptoms suffered by the victims of pesticide spraying—"tremoring, loss of ability to fly, paralysis and convulsions." Carson also wrote about household pets experiencing severe diarrhea and vomiting. Her moral outrage at such cruelty led her to ask, "By acquiescing in an act that causes such suffering in a living creature, who among us is not diminished as a human being?"[6]

Carson's advocacy of humane treatment for animals extended to her active participation in organizations that promoted conservation and animal rights. She was always an avid bird-watcher and active in her local Audubon Society. Members of that chapter provided her some important leads when she began researching *Silent Spring*. A letter Carson wrote to the *Washington Post* lamenting "the silencing of song birds" prompted a response from Christine Stevens, director of the Animal Welfare Institute (AWI), an animal-rights advocacy group. In seeking to ban or at least restrict medical research on animals, AWI encountered the same biases Carson confronted in writing *Silent Spring*. Many members of the science establishment dismissed antivivisectionists as the lunatic fringe of the animal-rights movement. Carson committed herself to Stevens's work and served as a member of the AWI advisory board. In her last years, she worked with Defenders of Wildlife to promote federal protections for research animals.[7]

Through her writing and her advocacy Carson placed herself at the center of a network of professional women, journalists, and club members who, though diverse in background, were united in their commitment to environmental protection and in their support for other women who shared their sympathies. Few of these women defined themselves as feminists. After all, *Silent Spring* preceded Betty Friedan's groundbreaking book *The Feminine Mystique* (1963) by a year. Carson stressed that she wanted to be championed for her message and her expertise, not her gender. Yet, throughout her career, strong women—Maria Carson, Mary Scott Skinker, Shirley Briggs, Marie Rodell, Marjorie Spock, Agnes Meyers—provided her with friendship, emotional and professional support, and intellectual affirmation. Women's organizations such as the American Association of University Women and the National Council of Women honored her and offered sympathetic audiences for her ideas. In turn, as historian Vera Norwood observed, "In Carson, her female colleagues found a spokesperson for their most deeply held convictions. Her symbolic power was increased by her gender. As important as Carson's message was the fact that a woman was the messenger."[8]

Norwood offered an intriguing explanation for why Carson's writing had such strong appeal to women. Ecologists often use as a metaphor for nature the concepts of "household" or "home," a "niche" or a "place." In writing about the oceans, Norwood observed, Carson described the life cycles of animals and their communities. Where then did these creatures, many of them adrift in the ocean currents, make their homes? With her capacity for close observation, Carson provided an answer, as Norwood suggested, that reassured "a generation of women raised to attach a great deal of importance to their homes as similar places of shelter." Ocean homes were central to Carson's descriptions of the interdependence of living things. Take, for example, the lowly barnacle. Its death, whether from attacking fish, worms, or snails or from other causes, left its empty shell attached to the rocks. Soon, that home welcomed new life—baby periwinkles or tide pool insects or, perhaps, "young anemones, tube worms, or even new generations of barnacles." Just as in the human world, Carson portrayed this shell house "as a haven from a threatening world. . . ."[9]

The more I learned about Carson—the environmentalist, career woman, scientist, animal-rights activist, faithful daughter and beloved aunt, female role model, devoted friend, gifted writer, and political dissenter—the more interesting she became. Her struggle to write *Silent Spring* while fighting against cancer was inspirational. The battle she waged against the male-dominated science establishment was truly heroic. In the introduction to the edition of *Silent Spring* my students read, former vice president Al Gore wrote, "Rachel Carson was one of the reasons I became so conscious of the environment and so involved with environmental issues." Just as Carson inspired him to write *Earth in the Balance*, she inspired me to write *The Gentle Subversive*.[10]

Anyone seeking to write a biography of Rachel Carson has one critical advantage—she was a writer herself. That means she left extensive papers from which scholars could reconstruct her life and work. She wrote thousands of letters and kept copies of most of them and the replies she received. On her research trips she took extensive field notes and preserved detailed records of her observations. Her manuscripts and early drafts of her books are all available, along with her childhood stories and college essays.

The voluminous correspondence between Carson and Dorothy Freeman, edited by Freeman's granddaughter Martha, has been made available in a published collection, *Always, Rachel.* One remarkable aspect of this friendship is that the two women actually spent little time together. Theirs was a friendship of words. Over the eleven years they knew each other, they were probably together no more than three or four months. Most of what they felt and thought they expressed in their letters and phone calls. Still, as with any subject, the record is hardly complete. Along the way some important materials were lost, including letters between Rachel and Dorothy that the two of them threw away and others that Robert Carson destroyed.

Any Carson biography faces another problem, one of daunting aspect: Carson was in a sense her own biographer. That does not mean she wrote extensively about herself or routinely revealed her inner thoughts and feelings. Quite the opposite. Carson remained an intensely private person who shared herself only with close friends and family, some of whom have added their own memories to the historical record. Rather, Carson explained what she cared most about in her books. And she did so with an eloquence that evolved out of years of dedication to her craft. There was almost no period in her life when Carson was not writing or producing books. Both admirers and detractors spoke of a "poetic" quality in her writing. Detractors suggested that by adopting the figurative language of the poet Carson deprived her prose of scientific accuracy and precision. Her admirers believed poetry allowed her to translate opaque scientific jargon into language any reader could understand, while at the same time evoking the wonders and mysteries of nature lost to scientists peering at the world through the lens of a microscope. Anyone who wants to know Rachel Carson can best do so by reading her books.

After encountering such eloquence firsthand, a reader will want to know more about the person who created it. That was certainly true for me, and that led to another problem, actually to two problems. Carson already has two major biographers, and others have written extensively about her. Paul Brooks, Carson's editor at Houghton Mifflin, first sought to honor her as one of his most successful authors and dearest friends. In doing so, Brooks had no desire to lift her veil of privacy. Instead, in *The House of Life: Rachel Carson at Work* (1972) he illuminated her life partly through biographical details that revealed her approach to writing and even more through numerous excerpts from her books and articles. For Brooks, what defined the poet in Carson was not so much her capacity for figurative speech as her gift for finding just the right word. That quality, when wedded to her "meticulous research," established, in his mind, a fruitful union between literature and science.

Brooks's reluctance to expose the private Carson left room for a biographer who intended to portray her full life. In *Rachel Carson: Witness for Nature* (1997), Linda Lear achieved that goal. Lear spent a decade learning all there was to know about her subject. She interviewed surviving friends, family, and work associates; visited a host of archives; scanned old newspapers; and tracked downs odds and ends wherever she could find them. There is little we are ever likely to discover about Carson that Lear has not already found. The biography she produced earned the praise of professional historians and Carson's friends and admirers. In it, Lear created a portrait of a person who treasured privacy and solitude yet who "wrote a revolutionary book in terms that were acceptable to a middle class emerging from the lethargy of postwar affluence and woke them to their neglected responsibilities."[11]

So why would we need another book about Rachel Carson? For one thing, no biographer, even one as skillful and thorough as Lear, can answer all the questions readers might have or anticipate all the issues a subject might raise. Lear could not know, though it would not surprise her, that critics in the twenty-first century would renew the attack on Carson's work and reputation. There are also some questions biographers choose not to address or for which they have no definitive answer. For some time now, reading audiences have been preoccupied with the sexuality of public figures. Was Abraham Lincoln gay, as a recent biography suggested? Did Thomas Jefferson have a slave mistress who bore him a child? Modern biography abounds with speculation on sexuality from the psychoanalytic to the political and to the prurient.

In this regard, Rachel Carson would seem to be a prime subject. What, after all, was the nature of her relationship with Dorothy Freeman? Readers of *The Gentle Subversive* must have noticed that I have said little about that subject. Neither did Linda Lear, despite her wide-ranging exploration of Carson's life. Both of us have chosen not to speculate whether that relationship (or the relationships with Grace Croff and Mary Scott Skinker at Pennsylvania College for Women) was sexual or platonic. The fact is that we can never know. More important, to reveal the essence of Rachel Carson, it doesn't really matter. Simply put, for most of her life, Carson devoted her emotional energies to writing, work, and family. When she met Dorothy, she found a kindred spirit, someone to whom she could unburden her heart, whose praise and affection meant more than the adoration of a demanding mother, who shared her devotion to the sea, who understood the passions aroused by the music of "LB" (composer/conductor Leonard Bernstein), and who was "a haven in a hostile world." Those feelings would have existed no matter what physical intimacy the two women shared.

Along with her many contributions to our understanding of Carson, Lear often corrected popular misconceptions about Carson's life and work. Al Gore reflected one such misreading when he wrote, "*Silent Spring* was born when [Carson] received a letter from a woman named Olga Huckins in Duxbury, Massachusetts telling her that DDT was killing birds." Paul Brooks similarly explained the origin of *Silent Spring* as a result of that letter from Huckins. Indeed, he considered the letter to have been sufficiently instrumental to include the entire text in his book. Carson herself fostered the impression that Huckins had inspired her crusade. Upon receiving Huckins's letter in 1958, Carson recalled that she "began to ask around for the information she wanted." The more she learned, "the more appalled I became, and realized that here was the material for a book." Shortly after she published *Silent Spring*, Carson wrote her friend Huckins, "It was not just the copy of your letter to the newspaper but your personal letter to me that started it all." In trying to find people to help Huckins, Carson discovered "that I must write the book."[12]

What better authority could a historian have than a subject's written testimony? Why doubt Carson's memory, especially since her editor also accepted the story? Well, because the story, even if logical and seemingly authentic, is not quite accurate, as Lear has demonstrated. In her forward to *Silent Spring*, Carson actually said that Huckins's letter "brought my attention back to a problem *with which I had been long concerned*" (italics mine). Indeed, she was already monitoring the Long Island spraying case and was aware of the plans for the fire ant eradication program. On January 27, 1958, the day Huckins sent her letter to the *Boston Herald*, Carson wrote to Marie Rodell for more information on spraying and to DeWitt Wallace at *Reader's Digest* to express her concern about pesticide programs.[13]

This may seem to be a small matter. So what if Carson exaggerated her debt in order to express gratitude to a friend who shared her concerns before most people ever thought about pesticide poisoning? Yet, the point is not inconsequential. The Huckins story, as Gore and Brooks relate it, suggests that *Silent Spring* began almost by happentance. Carson received a letter from a worried friend; seeking to help out, she unearthed a scandal and wrote a book that changed the world. The story, as Lear told it, is far more complicated. In 1957, Carson was looking for a focus for a new book. She had already encountered the DDT issue in 1945 and rediscovered it that summer. In Lear's version, *Silent Spring* would have been written even if Carson had never received a letter from Huckins. A complex series of events brought Carson to her subject and other external events, such as the strontium-90 scare, made her charges about DDT more credible. If there is a principle working here, it is not that great things come from small events but, rather, that ideas, like plants, grow best in well-cultivated soil.

Carson's explanation of her passion for the sea also complicates a biographer's efforts to understand this subject. Both Lear and Brooks mention the stormy night at Pennsylvania College for Women when Carson sat alone with Tennyson's poem "Locksley Hall." She later recalled that one line in the poem, "For the mighty wind rises, roaring seaward, and I go," seemed to predict "that my own destiny was somehow linked to the sea." Lear concluded, "Something in the moment clarified her direction, telling her that the 'vision splendid' she pursued lay with the sea." Brooks treats it more simply as a powerful memory.[14]

Many adults have a tendency to find a destiny in their lives arising from a single, powerful memory. Certainly, the tension at the center of the poem between the outer storm and the narrator's inner turmoil reflected Carson's sense that she faced a life-defining choice. Would she be faithful to her mother and the dream they shared of her career as writer, or would she shift her loyalty to Mary Scott Skinker and a more uncertain career in science? As Carson scholar William Howarth put it, "The poem also represented the end of childhood, because choosing science over literature was an act of obedience [to her mentor, Skinker] and defiance [to her mother]." That dark night spent with "Locksley Hall" most certainly touched powerful emotions, even helped clarify her decision between science and literature; but in the end, she chose both.[15]

I did not decide to write a new biography because I believed that Brooks and Lear had left important things unsaid about Rachel Carson. Indeed, there is little in either biography with which I disagree and much that I admire. My footnotes indicate how often I have relied on them for important information. Rather, I wanted a narrower focus that would allow me to emphasize how Carson and *Silent Spring* helped to inspire what Donald Worster called "the age of ecology." Further, I wanted to understand why, long after her death, some critics continued to attack her work and her reputation. And finally, because my book appears in Oxford's New Narratives in American History series, I hoped to recreate Carson's life as a story, albeit an instructive one.

Narrative, I realized, differs from traditional academic writing because it flourishes with the inclusion of vivid details. That is, a narrative writer seeks to include materials that scholars might dismiss as trivial. A scholarly biographer might write, for example, "The heat and humidity of that Washington summer day overwhelmed Carson." A narrative biographer would seek to make the reader feel the heat and humidity: "By noon of a sweltering summer day in Washington, Carson sat slumped in her chair mopping her brow. Her dress clung to the seat. She opened every window in hopes of capturing even the hint of a breeze." Searching for such details sometimes pays unexpected dividends. That became clear to me when I tried to describe the fire that I at first casually described as burning on the Cuyahoga River in 1969. I realized I actually knew almost nothing about it. Like other environmental historians, I had been content to marvel at the notion of a river in flames. Indeed, that is why the river became an oxymoronic icon of the industrial abuse of the environment.

The reality, it turns out, was considerably more complex, as my search for "details" soon revealed. Indeed, the Cuyahoga River could as easily have become a symbol of the new environmental movement. By the time it became a national sensation, the city of Cleveland and the state of Ohio had already undertaken projects to return the river to some semblance of health. Fish, an important sign of renewal, had returned to the lower reaches of the river at the time of the fire. Eventually, the federal government included a stretch of the river in the Cuyahoga Valley National Park not far from Cleveland.

More curiously, the fire turned out to be a virtual nonevent. Firefighters extinguished it even before reporters arrived on the scene so that the story barely made the back pages of the local papers the following day. Besides, the era of serious river fires ended well before this symbolic conflagration. The 1969 fire followed far more dramatic blazes in 1868, 1883, 1887, 1912, 1922, 1936, 1941, 1948, and—the most devastating of all—1952, a fire that resulted in nearly $1.5 million in damage. Thus, when *Time* magazine wanted an illustration to accompany its August story on the fire, it was forced to use a photograph, not of the event it was describing in 1969, but of the one from 1952. Nor was Cleveland the only city with a burning river. Pollutants fueled river fires in Baltimore, Buffalo, and Detroit.

Timing, more than the fire itself, made the Cuyahoga a symbol of environmental abuse. In 1969, the nation stood on the

This photo of a Cuyahoga River fire dates from 1952. That event inflicted far more damage than the 1969 fire. Firefighters extinguished the 1969 blaze before photographers arrived on the scene. (SOURCE: © Bettman/ CORBIS, Courtesy of Cleveland Public Library.)

brink of Earth Day and a new era of federal environmental activism in response to growing public demands for cleaner air and water. Cleveland was in the forefront of this new environmental concern. The city established the Cuyahoga River Basin Water Quality Committee in 1963 to reduce pollution, encouraged local industries to curtail discharges, passed a $100 million bond issue in 1968 to clean up the water (at the time, the federal government spent little more for the entire nation), and in early 1969 established the Clean Water Task Force to periodically sweep the river and to collect oil and debris.[16]

It was law professor Jonathan Adler who first took the trouble to explode this well-established myth. I wondered what inspired him to do so. One obvious explanation suggested itself—Adler teaches at Case Western Reserve University in Cleveland. This was obviously a hometown story. Yet, thousands of other academics and journalists in the Cleveland area ignored it for over thirty years. I also sensed something in the way Adler told his story that hinted he had a motive other than burnishing Cleveland's reputation. Then I noticed that Adler had written an earlier article highly critical of environmentalists, especially Robert Kennedy, Jr., a vocal advocate for the cleanup and protection of the Hudson River. Adler regretted that the Cuyahoga fire had inspired the drive for environmental protection at the national level. His point was not just that the fire was insignificant but, rather, that a correct reading of the Cuyahoga story did not justify the subsequent cleanup initiatives taken by the EPA or other federal agencies under the Clean Water Act. Adler argued that local and state environmental initiatives were a cheaper and more effective way to restore the river to health. "There is no reason to believe," he concluded, "that the adoption of federal command-and-control regulations was the only means of providing the level of environmental protection demanded by an ecologically awakened public."[17]

My determination to learn the height of the flames on the Cuyahoga in June 1969 (fifty feet) led me into this politically charged controversy. Was I now convinced that the fire was a myth and that I should remove it from my story or report it in another way to make quite another point? Not really. Cleveland may have spent $100 million in an effort to clean up its river prior to 1969, but it was still filthy and it did catch fire. Even if local reporters missed the story, a national audience did not. Nor does it matter much that the fire of 1952 was worse or that Baltimore, Buffalo, and Detroit had fires of their own. The fact remained that in 1969 the nation's rivers and air were polluted and most Americans wanted something done about it. In history, as in comedy, "timing is everything" but so are the details. Sometimes in getting the facts of the story straight, as Lear did with the Olga Huckins letter, historians uncover a new story. Adler sought to explode the Cuyahoga River myth for much the same reason conservatives renewed their attack on Rachel Carson—to discredit the EPA and the rationale behind federal regulation of the environment.

While narrative demands more details, a short biography covers less of the subject's life. The author must be selective. Lear took almost 500 pages to render Carson's life in full; I wanted a book about one-third that length. In the end, Rachel Carson solved my problem for me. She sometimes commented that while it was no mystery how she discovered her love of nature, she did not know how she became a writer. That would be my subject—Carson the writer. By looking at her four major books I could trace the evolution of her literary career from girlhood nature stories to *Silent Spring* and the rise of the modern environmental movement.

How then would I cope with the restraints that narrative imposes on historians, who generally use a more detached and analytical style of writing? Once again, Carson provided guidelines: effective narrative, as she created it, requires attention to the storyteller's point of view. Historians generally adopt a third-person, Olympian perspective, standing outside the narrative except when they enter to comment on its meaning. But in writing *Under the Sea-Wind*, Carson chose to tell her story from the point of view of the fish and other sea creatures whose world she wished to explore: "their world must be portrayed as it looks and feels to them—and the narrator must not come into the story or appear to express an opinion." In that way she could avoid the "human bias" that colored most stories about the sea. And by telling her story from the fish's perspective she gave it power and immediacy. Through the details of her subjects' life cycles, she brought her readers into their world. So it would be for me. I would tell Carson's story, as best I could, from her point of view, keeping my own commentary to a minimum. And because of the richness of the record she left, I seldom had difficulty showing how she thought or felt about her work and ideas.

Often, I found myself struggling to stay out of the story. For example, in Chapter Four, as I laid out Carson's research program for *Silent Spring*, I was struck by how much the debate her book triggered paralleled the current debate over global warming.

Through the research of Malcolm Hargraves, Morton Biskind, and Wilhelm Hueper, Carson obtained powerful circumstantial evidence about the carcinogenic threat posed by pesticides but without the proverbial smoking gun. To this day, the relationship between cancer and environmental pollutants remains contested. The same holds true for scientists wrestling with the issue of global warming. They have mountains of evidence linking greenhouse gases to rising temperatures but cannot make their case with absolute certainty. Many politicians and industrialists responsible for carbon dioxide emissions dispute their claims. The similarity between the two controversies struck me as a point worth making. But to make it in the midst of telling Carson's story seemed distracting. I wanted my reader to focus on her efforts to link DDT to cancer. Global warming was a matter for another time and place.

Something similar happened when I discussed Dr. Robert White-Stevens and his efforts to discredit Carson's case against DDT. One of his most telling points was that "Chemicals offer the only immediate hope of increasing food production to meet world needs." Without pesticides and fertilizers, the developing world faced growing hunger and disease. Those seeking to dismantle environmental regulations made the same case forty years later. They put Carson back on the stand and accused her of murder. But to forewarn readers that the issue would be revived forty years later was to get ahead of my story. Besides, if the point was clear enough, readers could make the analogy for themselves.

I struggled, too, in deciding how to treat properly the charge often leveled against Carson and the environmental movement that they practiced a misguided form of religion. Many of

Carson's admirers also granted the religious underpinnings of modern environmental thought. But where critics dismissed this faith as antiscience or paganism, historians such as William Cronon suggested that in the manner of what we call "religion," environmentalism "offers a complex series of moral imperatives for ethical action, and judges human conduct accordingly." If religion helps us to make sense of our lives in a world of infinite mystery and complexity, then Carson's approach to the natural world qualified as religious. Philosopher William James described religion as the "belief that there is an unseen order, and that our supreme good lies in harmoniously adjusting ourselves thereto." In that spirit, Carson challenged parents to direct children away from the "sterile preoccupation with things that are artificial, the alienation from the sources of our strength" and toward the wonders of nature.[18]

Certainly, Carson's writing contains the prophetic tone that Michael Crichton, for one, derided. For Cronon, by contrast, the link to the Judeo-Christian prophetic tradition has been an essential element of environmentalism that offers "predictions of future disaster as a platform for critiquing the moral failings of our lives in the present." Thus, unlike most political movements, environmentalism is openly moral in its prescriptions for our daily lives. As Cronon observed, it "touches every personal choice or action." A brick in the toilet tank, a hybrid car, organic foods—these are moral as well as practical choices.

Once I recognized the moral underpinnings of Carson's work, I could better understand why her critics considered her a subversive. She did not simply challenge her readers to demand tighter regulation of DDT, though she hoped they would. Even more, she wanted them to understand that human actions affect the natural systems upon which all life depends. The destruction of nature had moral as well as physical consequences. And when people lost touch with what was natural, she believed they lost touch with what made them human. Religion and science came together at the end of her 1955 book *The Edge of the Sea*. She asked her readers to wonder "what is the meaning of so tiny a being as the transparent wisp of protoplasm that is sea lace, existing for some reason that is inscrutable to us—a reason that demands its presence by the trillion amid the rocks and weeds of the shore?" To which she concluded, "The meaning haunts and ever eludes us, and in its very pursuit we approach the ultimate mystery of life itself."[19]

These lines must have expressed her deepest conviction because Carson asked to have them read at her funeral. Equally significant, her brother Robert refused that request. Somehow they did not suit the formal service he planned at the National Cathedral. All the same, the friends who gathered to remember her at All Souls Unitarian Church did hear them read. They were, I came to believe, the words of a prophet, though a secular one. A traditional Christian such as Robert Carson must have found his sister's convictions a threat to his beliefs. But for those willing to open their hearts and minds to the mysteries of the living world, to be humble in the face of nature, and to welcome the surprises nature offers to those who take time to look and listen, Carson offered a path to a higher spiritual plane.

Many prophets are controversial in their own lifetimes because they challenge the orthodoxies of their day. That may be why those in authority so often dismiss them as subversives.

Rachel Carson grew up in an America whose leaders preached that the nation had achieved greatness by conquering and controlling nature. How then could they accept a woman who challenged them to be respectful of nature, to recognize the limits of their own understanding, and to seek out the mystery and wonder of the unknown? They were, after all, practical men. Rachel Carson urged them also to be men of vision.

Notes

1. Worster, *Nature's Economy*, p. xiii.
2. William Cronon, *Changes in the Land: Indians, Colonists, and the Ecology of New England* (New York, Hill and Wang, 1983), pp. 6–8.
3. Carolyn Merchant, *Ecological Revolutions: Nature, Gender, and Science in New England* (Chapel Hill, University of North Carolina Press, 1989), pp. 17–19.
4. Paul Shepard, "Introduction: Ecology and Man—A Viewpoint," in Paul Shepard and Daniel McKinley (eds.) *The Subversive Science: Essays Toward an Ecology of Man* (Boston, Houghton Mifflin, 1969), pp. 1–10. The best statement about misguided technology and ecological crisis is Barry Commoner, *The Closing Circle: Nature, Man & Technology* (New York, Random House, 1971).
5. Haney quoted in Brooks, *House of Life*, p. 264.
6. Carson, *Silent Spring*, pp. 90, 100.
7. Norwood, *Made from This Earth*, pp. 160–163.
8. Ibid., p. 164.
9. Ibid., p. 150; see also Carson, *Edge of the Sea*, p. 55.
10. Al Gore, "Introduction," in Carson, *Silent Spring*, (Boston, Houghton Mifflin, 2002), p. xviii.
11. Lear, *Rachel Carson*, p. 5.
12. Al Gore, "Introduction," in Carson, *Silent Spring*, p. xviii; Brooks, *House of Life*, pp. 233–236.
13. Lear, *Rachel Carson*, pp. 315–316, 422, fn. 14, p. 546 and fn. 89, 90, p. 571.
14. Ibid., p. 40; Brooks, *Rachel Carson*, pp. 20–21.

15. Howarth, "Turning the Tide," p. 44.

16. www.pratie.blogspot.com/2005/03/cuyahoga-river-fire-of-1969.html. See also Jonathan Adler, "Smoking Out the Cuyahoga Fire Fable: Smoke and Mirrors Surrounding Cleveland," June 22, 2004, *National Review* online, www.Nationalreview.com.

17. Adler, "Smoking Out the Cuyahoga."

18. I had the good fortune while I was writing the afterword to discover that Thomas Dunlap, who wrote *DDT* and several other important books on environmental topics, had just published *Faith in Nature: Environmentalism as Religious Quest* with an introduction by William Cronon. See, for example, pp. 3–13 to see how they reconsider what Michael Crichton dismissed as a pseudoreligion.

19. Carson, *The Edge of the Sea*, p. 216. That vision led Katherine Moore to write "The Truth of Barnacles: Rachel Carson and the Moral Significance of Wonder," *Environmental Ethics*, Fall 2005, Volume 27, Number 3, pp. 265–278.

BIBLIOGRAPHY

BOOKS AND ARTICLES BY RACHEL CARSON

"Undersea," *Atlantic Monthly* 160 (September 1937), pp. 322–325.

Under the Sea-Wind: A Naturalist's Picture of Ocean Life (New York, Simon and Schuster, 1941).

"Birth of an Island," *Yale Review* 40 (September 1950), pp. 112–126.

The Sea Around Us (New York, Oxford University Press, 1951).

The Edge of the Sea (Boston, Houghton Mifflin, 1955).

"Help Your Child to Wonder," *Woman's Home Companion* (July 1956), pp. 25–27, 46–48.

"Our Ever-Changing Shore," *Holiday* 24 (July 1958), pp. 71–120.

The Sea Around Us, rev. ed. (New York, Oxford University Press, 1961).

"Silent Spring" (excerpt) in "Reporter at Large," *The New Yorker* (June 1962), pp. 16, 23, 30.

Silent Spring (Boston, Houghton Mifflin, 1962).

"Rachel Carson Answers Her Critics," *Audubon* 65 (September/October 1963), pp. 262–265, 313–315.

The Sense of Wonder (New York, Harper & Row, 1965). Photographs by Charles Pratt.

Rachel Carson Papers. Yale Collection of American Literature, Beinecke Rare Book and Manuscript Library, Yale University.

Books and Articles about Rachel Carson

Brooks, Paul. *The House of Life: Rachel Carson at Work* (Boston, Houghton Mifflin, 1972).

Freeman, Martha, ed. *Always, Rachel: The Letters of Rachel Carson and Dorothy Freeman* (Boston, Beacon Press, 1995).

Graham, Frank, Jr. *Since* Silent Spring (New York, Fawcett, 1970).

Hazlett, Maril. " 'Woman vs. Man vs. Bugs': Gender and Popular Ecology in Early Reactions to Silent Spring," *Environmental History* 9 (October, 2004), pp. 701–729.

Howarth, William. "Turning the Tide: How Rachel Carson Became a Woman of Letters," *The American Scholar* 74, (Summer 2005), pp. 42–52.

Lear, Linda. *Rachel Carson: Witness for Nature* (New York, Henry Holt, 1997).

Murphy, Priscilla Coit. *What a Book Can Do: The Publication and Reception of* Silent Spring (Amherst, University of Massachusetts Press, 2005).

Norwood, Vera. *Made from This Earth: American Women and Nature* (Chapel Hill, University of North Carolina Press, 1993).

Rubin, Charles T. *The Green Crusade: Rethinking the Roots of Environmentalism* (New York, The Free Press, 1994).

Smith, Michael B. " 'Silence Miss Carson!' Science, Gender, and the Reception of *Silent Spring*," *Feminist Studies* 27 (Fall 2001), pp. 733–752.

Waddell, Craig. *And No Birds Sing: Rhetorical Analysis of Rachel Carson's* Silent Spring (Carbondale, University of Southern Illinois Press, 2000).

Environmental Histories

Cronon, William. *Uncommon Ground: Rethinking the Human Place in Nature* (New York, Norton, 1995).

Diamond, Irene, and Orenstein, Gloria Feman, eds. *Reweaving the World: The Emergence of Ecofeminism* (San Francisco, Sierra Club Books, 1990).

Dunlap, Thomas R. *DDT: Scientists, Citizens and Public Policy* (Princeton, NJ, Princeton University Press 1981).

————. *Faith in Nature: Environmentalism as Religious Quest* (Seattle, University of Washington Press, 2004).

Gottlieb, Robert. *Forcing the Spring: The Transformation of the American Environmental Movement* (Washington, D.C., Island Press, 1993).

Helvarg, David. *The War Against the Greens: The "Wise Use" Movement, the New Right, and Anti-Environmental Violence* (San Francisco, Sierra Club Books, 1994).

Karaim, Reed. "Not So Fast with the DDT," *The American Scholar* 74 (Summer 2005), pp. 53–59.

Lytle, Mark H. *America's Uncivil Wars: The Sixties Era from Elvis to the Fall of Richard Nixon* (New York, Oxford University Press, 2006).

Merchant, Carolyn. *Earthcare: Women and the Environment* (New York, Routledge, 1996).

Nash, Roderick. *Wilderness and the American Mind*, 3rd ed. (New Haven, Yale University Press, 1982).

Russell, Edmund. *War and Nature: Fighting Humans and Insects with Chemicals from World War I to* Silent Spring (New York, Cambridge University Press, 2001).

Shabecoff, Philip, *A Fierce Green Fire: The American Environmental Movement*, 2nd ed. (Washington, D.C., Island Press, 2003).

Stoll, Mark. *Protestantism, Capitalism, and Nature in America* (Albuquerque, University of New Mexico Press, 1997).

Worster, Donald. *Nature's Economy: A History of Ecological Ideas*, 2nd ed. (New York, Cambridge University Press, 1994).

INDEX